Manufacturi[illegible] Management
Learni[illegible]gh Case Studies

Other titles of related interest

Design and Manufacture – An Integrated Approach
Rod Black

Form, Structure and Mechanism
M.J. French

Management of Engineering Projects
Richard Stone (Editor)

Mechanical Reliability, second edition
A.D.S. Carter

Reliability for Engineers – An Introduction
Michael Beasley

Manufacturing Management

Learning through Case Studies

D.G. Coward

First published 1998 by
MACMILLAN PRESS LTD
Houndmills, Basingstoke, Hampshire RG21 6XS
and London
Companies and representatives
throughout the world

ISBN 0–333–64777–7

A catalogue record for this book is available from the British Library.

This book is printed on paper suitable for recycling and made from fully managed and sustained forest sources.

10 9 8 7 6 5 4 3 2 1
07 06 05 04 03 02 01 00 99 98

Printed in Malaysia

To Sheila with Deborah, Andrew, Robert and Richard

Contents

Preface

Note to readers: please do not skip over this Preface or Chapter 1

For several years I have noted the weaknesses of many engineering text books, particularly their unsuitability for the young when coming across them for the first time. One weakness is centred on the application of theory into practice. Examples, when given, do not present enough substance when compared with real, practical problems. Budding engineers are confronted by text book solutions to problems that they do not know exist or why they exist.

These young people are unaware of the subtleties of course titles based on Industrial Engineering, Manufacturing Systems, Operational Management, Logistics and Production Management. They have to jump many hurdles in faith and trust before they are able to integrate their acquired knowledge of industrial problems into their experience. Only then can they make a contribution towards solving a firm's problems. Why not gain some of this insight earlier rather than later? It can be realised that the process of gaining academic knowledge is not always related to the real world, because many solutions in industry are non-mathematical. Often these are off-track compared with text book presentation: general versus specific. It follows that if a book could be constructed, giving practical examples, this would assist the young 'potential engineer' to seriously consider the merits of the engineering profession. Only then may a demand arise from young students, instead of rejecting engineering based on misleading information.

One area of engineering activity which can be the 'Cinderella' of a school career department's interest is Operational Management. For such a worthwhile challenging career it is unfortunate that there is too often a lack of experience and insight from career personnel.

Some consider this to be the reason why a classic discipline such as Mechanical Engineering is chosen and the production derivatives ignored. This circle of input and output needs breaking and counteracting with factual operational information. No doubt others further along their career path will also be interested to develop their learning by using such a book.

Case studies seem to be the best vehicle to deliver the necessary insights. However, many casebooks only present a problem for analysis. Few bother to discuss solutions unless a mathematical basis can be used. It was against this background that the specification for this book originated. Case studies based on the author's experience would be outlined with added details and comments about the particular theories applicable in practice. Readers need to be asked to contribute their ideas and analyse each case from directed questions. Their answers can then be compared with the actions taken by companies, which are reported at the end of each case. The final section also gives the actions that the management took, with the reasons behind the case particulars. When available, the outline results of the changes carried out are disclosed. Such is the flexibility of the case contents and sequence that merely reading through each section like an unfolding story would not take long and prove to be of great benefit.

It is considered that the formulation chosen will help student engineers to experience something of the investigation and improvement activities involved in the day-to-day running of a successful business.

There will be enough typical information supplied to allow each student reader to decide upon the degree of interest in the profession. The student can then direct his or her efforts towards relevant goals.

Young potential engineers could start reading these cases at any age above sixteen and later find them of use during their professional engineering training. This applies to their work both at university and at a company.

All kinds of experiences are essential, especially when combined with insights into the reasons for appropriate management decisions and actions. The experiences given, which are based on real-life situations, have been simplified, modified and coded to protect the companies involved.

The case studies are presented for use in discussion and learning processes and do not imply any criticism of the management's actions.

The Saltwards Group, along with company names and personnel, are all fictitious. However, the situations discussed are from the real-life experiences of the author; the fictitious framework and data protect confidentiality.

Finally, I want to thank my long-term colleague, John Salt, for his help in formulating this book and encouraging me to proceed. Unfortunately, John was so busy advising about problems in industry that he could not find time to make the valuable contribution that only he could offer.

David Coward

1 Introduction to the Casebook: Structure and Company Information

Why Use a Casebook Approach to Learning?

Textbooks are excellent vehicles for presenting theory and general principles, but they presuppose that the reader will be able to apply this knowledge to specific situations after studying only the smallest amount of example material. This leaves a large gap between stated knowledge and the techniques needed for success in the world of application and to satisfy the typical expectations of the lower-level line manager. The budding operations manager/engineer is left in the middle of this void and needs help.

Two professions that traditionally study cases are law and medicine, based upon the following reasons:

- there is a great deal of working knowledge not specifically laid down in law;
- problems presented in medicine by patients are not necessarily the true ones, since symptoms can be misleading.

Business operations are similar to both law and medicine in that these factors are ever present.

A typical casebook attempts to represent real-world situations for consideration and interpretation by the reader so that the learning process is given better insight. A gap still exists because there is no feedback to inform the reader either about relevant theory or the reasons for the specific solution adopted. Even then there is some further confusion because an open-ended situation is often presented for analysis and no solutions given.

The concept behind this casebook is to provide all the reader should reasonably need in relation to an industrial difficulty. The reader will be encouraged to give a positive response to the challenges built into each case.

Essentially the subject matter is presented in a critical style, the problem is described, relevant theory given and then the reader is able to make an individual contribution. This contribution can then be compared using the feedback loop provided, which quotes the actions that the company took together with reasons and the consequential results to the new changes. Each case included is based upon real-world situations. Hopefully, this extra information will bridge the gap between theory and practice, providing an important experience for prospective managers and engineers of all disciplines.

Each case study is strongly based on the experience of the author who made a significant contribution to the changes included. The material presented has been simplified for reasons of impact on the learning process and incorporates modified data to protect the companies involved. Every case has been selected to represent the range of operational activities important to the overall success of the company. They are set in areas such as Strategy, Production Control, Human Resources, 'Just in Time', Process Control, Costing, Investment and Layout situations. The comprehensive nature of this list demonstrates the widespread nature of operational sources of saving.

The Structure of All Case Studies

Each case is presented in a similar format to enable the reader to become proficient at systematically tracking through the necessary detail in order to be able to reread a particular section. To gain full advantage of the quality material presented, the reader should systematically analyse every step in a critical fashion and hence formulate ideas to answer the questions set. All the facts presented show the way forward to some of the conclusions being made.

The main headings of a standard format are as follows:

Sections A and B: Learning Outcomes and Learning Objectives
Specific listed outputs under each heading demonstrate the advantages and areas of personal development and insight that the reader should have gained after a detailed appraisal of the particular case.

Section C: The Basic Theory (of the specific content of the case)
This section, by its very nature, has to contain a large amount of material so that the theory can be outlined. Often the theories are critically examined to reveal their strengths and weaknesses. It is treated in this way to counter the traditional textbook approach which can be uncritical while presenting *all* theory as valid and useful. When this style is adopted it confuses many people, and this book is an attempt to counteract this tendency by providing clear guidance. Open-ended cases could have a wide range of theories discussed, so specific editing had to be used based on the premise that:

1. readers may have some prior knowledge;
2. Readers can use the references or other suitable texts to fill particular shortfalls in their experience.

Section D: Introduction to the Problems of the Company
The company is introduced and an outline of the main concerns of the management is expressed. A potted history is included if it helps to clarify the extent of the changes that may be needed. Some of these outlines have to cover many years because of the drift in performance that has occurred. Detailed information is given in reports from a task force team who have systematically obtained facts which will help to unravel the case complexities. The problems discovered provide a canvas on to which the modern theory and techniques can be painted, tested and evaluated.

Section E: Questions for the Reader and References – Clarification of Problems
Alternatives are an important feature of the management of change so the reader is encouraged to consider and evaluate as many alternatives as possible. Questions are set to aid this activity, with a list of some reference material which can be useful but is not essential to carry out the objectives of the case. Only those people who consider they are unable to proceed need look elsewhere.

Section F: Proposals Adopted by the Company
In this section details of changes carried out, with reasons, for the particular company provide the feedback to complete the reader's learning outcomes. Whenever available, performance figures resulting from the changes made are included to highlight the difference between past performance and the new conditions.

Finally a summary is given to pull together the different strands of the case and assess the overall effectiveness which can provide strategies and tactics for future actions.

Background to the Case Study Companies

All the cases are set in the same large group of companies, Saltward Group Plc. This international group has a central Management Services/Industrial Engineering facility. From this centre of excellence, highly skilled and also practical consultants act as 'trouble-shooters' in the business units attached to each of the main product areas (see Figure 1.1). The diverse nature of all these autonomous business units creates an ideal setting for a range of different but complementary case evaluations.

The main group board of Saltward audits the targets set for each Product Based Board and their constituent business units. When necessary the main group can provide capital and resources at advantageous rates to assist the growth and survival of each unit. However, failure to provide a plan with set targets and margins leading to little chance of achieving them in the

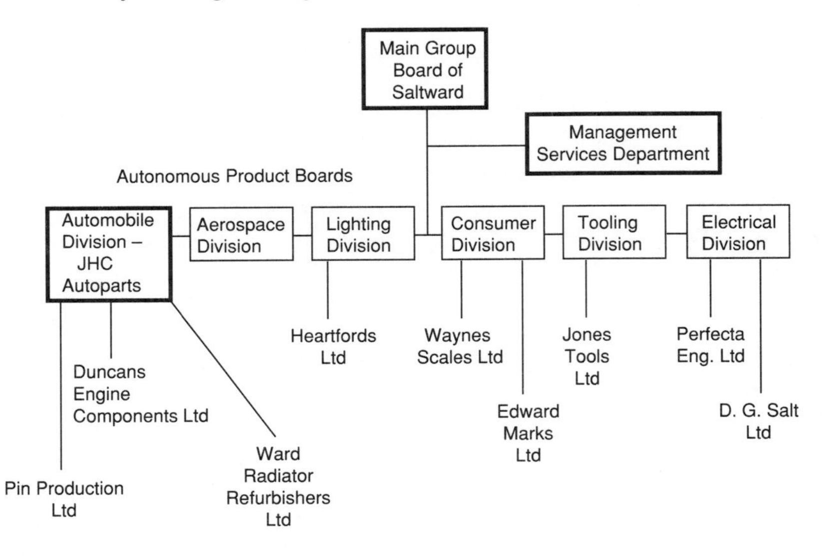

Figure 1.1 Companies involved in casebook, positioned in group – organisation structure of Saltward Group plc

near future can result in either the closing down or selling of the offending business unit.

This pressure for continuous improvement has only recently been evident and was one of the developments built into the design of the new organisation structure (Figure 1.1 and see Chapter 2). It is because of these strategic changes that the business units need to show some fundamental responses, however only a few of these can be included in this book.

The different firms that have a part in this casebook are extremely varied, so a short introduction of each is included to set the scene. Further details are given when a particular case is investigated.

The Decision Process

Each case study will involve making decisions. How are these decisions made? Here is a typical methodology that can be applied to many, if not all, situations. Decisions need to be made every day to keep a company working. Some of these decisions are more important than others and may have far-reaching effects on the company. Managers are commonly thought to make decisions as an integral part of their job description. How does a manager make such a decision? To try to understand the process look at the following example.

Requests have been received from Department A concerning the need to have further internal transport available. Two of the requests mention a fork lift truck. Suggest three stages that could be involved in the managers' decision process.

The general pattern of decision making may be observed in many different situations. A generic model was described by Herbert A. Simon in the 1960s. Simon's model divides the process into three stages. These stages involve:

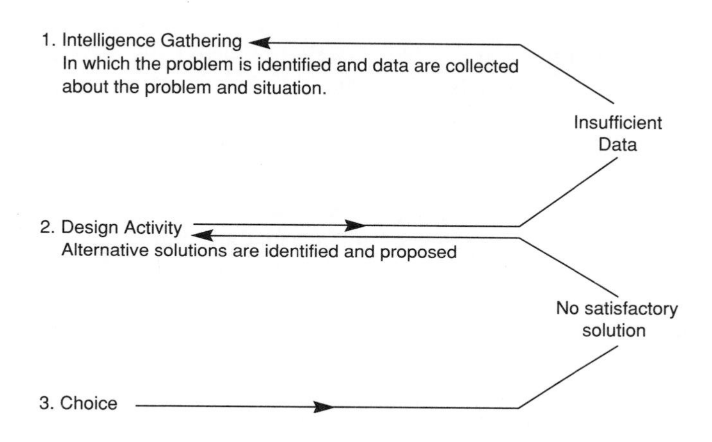

A solution is selected and its implementation monitored.

Notice that Simon includes iterative loops which enable the decision maker to return to earlier stages if:

(a) management does not have sufficient data to identify either the problem or a potential solution;
(b) the selected solution does not satisfactorily solve the problem.

Managers also have to make decisions which involve different timescales as well as having differing levels of impact on the company. A framework was first suggested by R.N. Anthony in the 1960s. The three different types of decisions have the following characteristics:

Operational Decisions

These decisions are involved in the day-to-day running of a company. The operational-level manager has a fairly standardised decision-making task, much of which can be automated, involving the main groups of primary activities: Sales Order Processing, Purchasing, Stock Control, Production Control, Warehousing and Distribution. Typically there is a large flow of detailed data, within and out of the organisation, via these routine activities – see Chapters 6, 9, and 10 for example.

Tactical Decisions

These decisions are concerned with short- to medium-term implications, and take account of the operating-level performance achieved, monitoring the use of resources to satisfy customers. Most data relevant to tactical decisions are derived from summaries of data used at the operating level. Major decision-making areas include stock policies (such as reorder levels), purchasing (such as choice of supplier), production (such as master production schedule), distribution and personnel decisions. Typically such decisions are made by departmental heads – see Chapters 3, 4, 11 for example.

Strategic Decisions

These decisions concentrate on a wider range of issues involving the medium- to long-term performance of the company, and are naturally associated with the future (for example, projections and a range of forecasts). External influences play an important role at this level of decision making. Major areas include determination of business aims, analysis of organisational performance, formulation of company policies and structure – see Chapters 2 and 5 for example.

The Companies Outlined

- **JHC Autoparts Group** **Chapter 2**

A subsidiary of Saltward to which dramatic changes are brought about as a consequence of a complete rethink on the strategy of the business. Without a complete reshaping of the group, the decline of many years subsidised by the rest of the main Saltward Group, will result inevitably in partial closure.

- **JHC Duncans Engine Components Ltd** **Chapter 2**

Part of the Autoparts sub-group which is established by the new strategy because it identified as a profit-making area with potential for growth. The firm is used as an example of the changes brought about internally to enable the strategy to operate and to discuss the outcomes and deliverables resulting soon after the completion of phase 2 of the plan.

- **Edward Marks Ltd** **Chapter 3**

A consumer division company that has recently had its managing director replaced. The new managing director wants to change strategy and management style. As he is new to the company he wants a comprehensive study of the present system and what options are open to him. Various organisational structures are considered with strengths and weaknesses investigated. As the management structure is large the scope for change is considerable. The reasons for the investigation are analysed and choice highlighted.

- **Jones Tools Ltd** **Chapter 4**

A relatively small company that manufactures press tool parts to customer designs and is having difficulty in delivering these parts on time. For many years the company had been successfully operating in a sellers' market which enabled its shortcomings to be hidden, and little customer reaction ensued. Now it faces more competition in a buyers' market – it is a case of improve considerably or business will dry up.

- **Waynes Scales Ltd** **Chapter 5**

Electronic weighing apparatus connected to computer processing is custom built for a variety of users in the distribution and retail outlets field. The company has been successful in recent years but delays on some major units are affecting the future performance and competitive power in the view of some of the management. Subcontract work is being used to fill this shortfall but the main company considers it is not in full control of this supply. A feasibility study and action on costing systems are proposed to see if a considerable load of work could be economically produced in-house.

- **Heartfords Ltd** **Chapters 6 and 7**

This comprises one of two sites situated 300 miles apart making lighting products for the general commercial market and for the specialist user.

The site was designed for a different purpose and is unsuitable in many respects, so any changes required by management need to keep this factor in mind. Output is not profitable so action is required to prevent the site closing, with work being transferred to the other. Major changes are undertaken in production systems and assembly lines to improve cash flow.

- **Perfecta Engineering Ltd** **Chapter 8**

An attempt to introduce AMT machining centres has had a mixed result in performance. However, upper management is confident that, by learning from its mistakes, AMT can be successful. Overcoming shopfloor inertia and making the investment pay are the tasks set before the consultants.

- **Ward Radiator Refurbishers Ltd** **Chapter 9**

This is a very small company obtained by Saltward as part of a takeover of more desirable companies. The two main managers are father and son, who look after a business subject to competition that it has failed to respond to. Vehicle radiators are refurbished by breaking them up, replacing defective components and reusing parts unaffected by past activity.

Output must be increased by reducing throughput time, which is mainly delay. An outside influence on change is necessary because internal attempts have failed.

- **Pin Production Ltd** **Chapter 10**

This is a small specialised producer of gudgeon pins for a few engine manufacturers, and employs 40 people. Quality assurance has become an essential part of company activity; this reveals capacity problems and too much internal scrap produced by one major process. Unless the quality and capacity can be increased then the firm will be made bankrupt by having to accept the cost of failure of this critical component. Initially the responsibility for change falls on the Quality Manager and his investigations reveal a complicated production problem.

- **D.G. Salt Ltd** **Chapter 11**

This is one of the companies that requires complete redesign as part of the group strategy of developing technological products. Part of these changes require the firm of 450 employees to move to another site. The budget is tight and time short because of the need to keep supplying parts to customers throughout the inevitable disruptions. A team is used to plan and control the move on a day-to-day basis with detailed activities highlighted by visual control. Results to all actions are recorded.

Level of Information

It is commonly assumed that the secret of effective decision making is information: the more information available, the better the decision made.

Managers need information so that they:

1. can effectively meet the objectives of the company's business plan;
2. effectively schedule the company's resources of money, machines and people – note that constraints exist in these areas;
3. can perform their managerial duties of planning, organising and controlling.

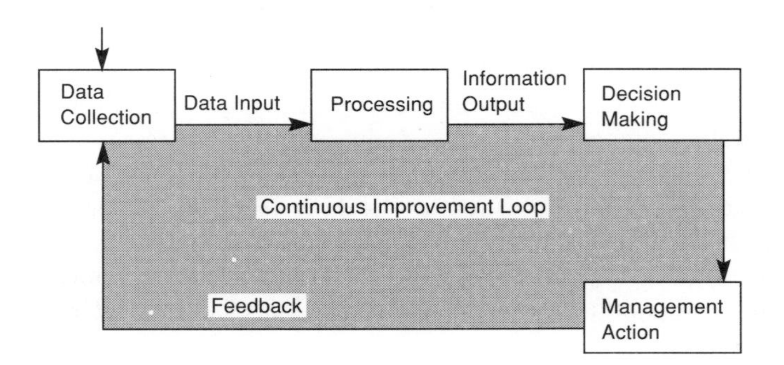

Figure 1.2 Elements of a management information system

sponsible for the overall operation of the business. The organisation is one of the major methods he employs to ensure the company's objectives and strategy. Finally, organisational design is not decided by democratic voting and what is preferred by most people; only one vote counts and this belongs to the Managing Director. This is one of the main reasons why each occupant of the top post invariably alters a company's structure in fundamental ways to suit the new situation.

A simplistic attitude is, therefore, to give managers all available information and generate more later. However, care is needed in the choice of information/data, because too much can be as bad as too little.

Notice how the different investigators in later chapters go about obtaining specific data and then turn it into useful information.

Theory to practice is moving from the general to the particular. This process is a real challenge and Figure 1.2 gives the theoretical side of the activity.

It is often known 'what needs to be done' but as always it is 'how to do it' that is the challenge. It is good use of data turned into effective practice which leads to success.

2 Strategy Formulation of the Group

Section

A. Learning Outcomes

Change is a fact of life and some consider that it is best achieved through systematic analysis of need, capability to meet the need, resources available and timescale required to effect change. By having a plan to control change it is more likely that success will be achieved. Perfection is not sought or expected and so changes to the plan and the objectives set are a normal occurrence. However, the reasons for the change of plan are from analysis and this helps to keep strategy realistic. The alternatives of 'hope and see', 'fly by the seat of your pants' or 'do your best on a daily basis and the future takes care of itself' are not recommended except for policies of despair.

This case study will demonstrate that:

1. Past policies used – such as stable design and production methods, relying on a steady demand for sales, competition is not to be feared, quality

and reliability of products are adequate, investment can only be based on present profits not on potential profits – are the forerunners of failure.
2. Data collection and analysis are essential features of modern systems of decision making. The idea that the company already collects all the information needed is flawed. New methods of collection and analysis based on direct observation are often the only reliable source of information.
3. Benchmarking the main company performance measures, then comparing them with achievements world-wide, can be the catalyst for rapid strategy changes to gain parity.
4. Once the whole outline strategy has been laid, the full realisation is achieved by each business unit, department and individual playing their part. Just like bricks in a wall, each is important but so is the overall strategy that 'cements' them together.
5. The two strategies of 'top down' and 'bottom up' each have a role to play, one supports the other. Too often only 'top down' has been used, with little success.
6. An overview of the essentials of management action is a useful study but the actual activity of being involved in a situation like this one is invaluable experience.

B. Learning Objectives

After studying and analysing this case study the reader should be able to:

1. Explain the steps in strategy formulation.
2. Identify key activities in strategy formulation.
3. Define organisational philosophy.
4. Define organisational culture and mission.
5. Differentiate between long- and short-range objectives.
6. Identify strategic business units.

C. The Basics of Strategy Formulation

Some companies grow and become profitable, while others are taken over or go bankrupt. Some companies expand into new markets and diversify their product range while others do not. Giants are being created each decade and some of the older giants disappear or become average or small. For example MacDonalds, Amstrad and Hanson have grown, whereas Triumph Motor Cycles, Austin and Morris Cars, and Tube Investments have contracted or even disappeared.

In years to come the increased rate of social, environmental and technological change, the increased multinational spread into a world marketplace along with limited resources will make the competition of organisations even more complex.

How do companies make decisions about their future with this background? The method used is called *Strategic Management*. This technique is equally useful to any organisation whether in the field of manufacture, services, public institutions or public utilities.

Strategic formulation is concerned with making decisions about:

1. Defining a company philosophy and mission.
2. Establishing long- and short-range objectives to achieve the mission.
3. Selecting an appropriate feasible strategy to achieve the objectives.
4. Having a *strategy implementation*, that is the plans and controls at all levels to produce success.

Putting company philosophy into writing a 'mission statement' is an important feature of systematic analysis, the written word being more revealing than the spoken one. Peters and Waterman[1] found that effective companies have the following beliefs built into their philosophy:

(a) belief in being the best;
(b) belief in the importance of the details of execution (the nuts and bolts of doing the job well);
(c) belief in the importance of people as individuals;
(d) belief in superior quality and service;
(e) belief that most members of the organisation should be innovators, and its corollary, the willingness to support failure;
(f) belief in the importance of informality to enhance communication;
(g) belief in a recognition of the importance of economic growth and profits.

A 'hierarchy of strategies', explained later in more detail, can be shown on a top down basis as in Figure 2.1.

No attempt is made here to give detailed procedures on strategy formulation but just to raise total awareness in this important area of planning and control.

The importance of systematic formulation to give feasibility is now outlined so it can be used in the process of analysis and consideration of the alternatives. The emphasis is on feasibility, that is, can the objectives of the company be achieved within the timescale and total resources of the company?

Below is one suggested structure or methodology; many variations exist but they should always include the main steps highlighted to be successful.

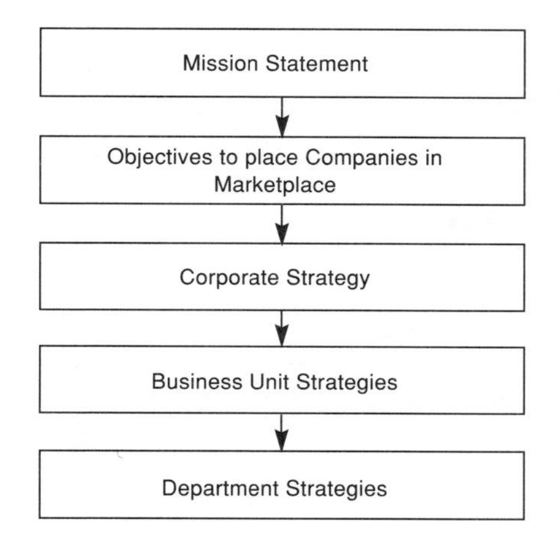

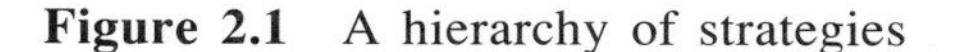

Figure 2.1 A hierarchy of strategies

Strategies to close the 'gap' between the status quo *and company needs*

A way of considering the severity of the change in strategy needed is to analyse various indicators of activity, such as sales, employment requirements, product range etc.

The reason for the analysis is to contrast the position of the company over a period in the future and determine:

(a) Where the company would be if it did nothing new.
(b) Where the company management wants to be as a well-established company.

This is easily shown on a simple graph (see Figures 2.2 and 2.3). The features of these graphs are that they indicate two different scenarios which will need two different responses in terms of the degree of change to be planned and obtained.

Scenario 'XZ' (Figure 2.2)
A rapid decline faces the company and the strategy used must encompass a large *gap* between the present policies and where the management thinks it needs to be. Hence radical changes across the company are needed to close the gap. This represents the conditions of the case study under consideration.

Scenario 'YZ' (Figure 2.3)
Here the policies being used at present will produce a rising response but it is not as high as management wants. The gap is relatively small and only fine-tuning to the present strategy is required to obtain the necessary result.

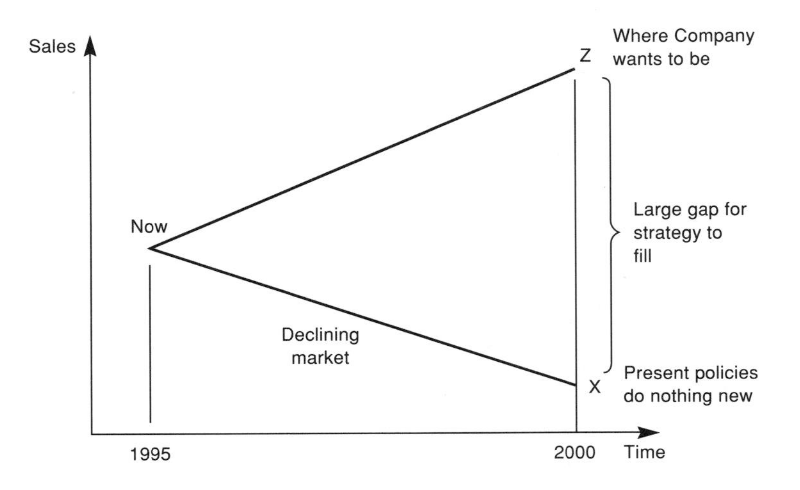

Figure 2.2 Strategy analysis in a declining market

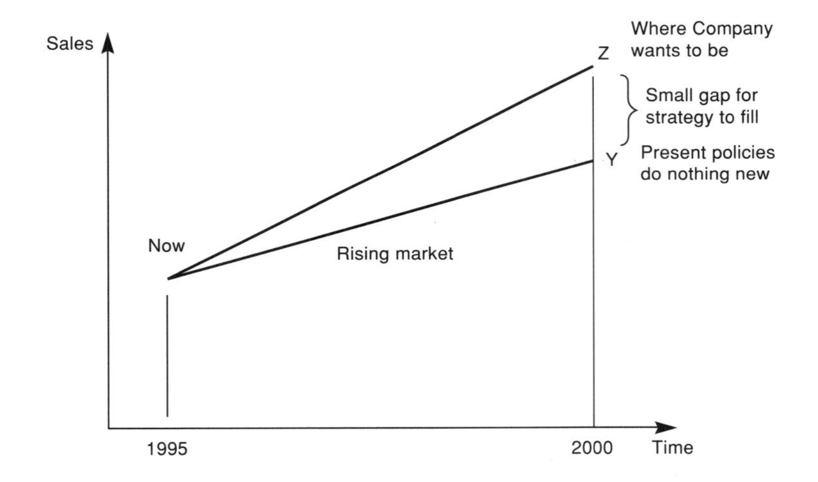

Figure 2.3 Strategy analysis in a rising market

This process is very basic but it concentrates the mind on the real problems faced by setting them in a framework in which the discipline of the policy 'doing nothing new' are evaluated. This is a challenge for some managements because the present problems can seem so large and immediate that future plans and thoughts are relegated by the apathy of 'I will when I have time'.

D. Outline of Strategy Formulation

An interactive process of formulation is given by the following stages:

Know the marketplace	Is this the business we want to be in?
↕	
Set objectives	Targets that must be met
↕	
Consider policies	Plans to achieve objectives
↕	
Forecast the effect on resources	Is there enough capacity?
↕	
Budgets	Is the cash flow pattern acceptable?

These stages are given with two-way arrows to emphasise the balance of feasibility attributes required to make a homogeneous strategy. This demonstrates the movement up and down between stages that is necessary to ensure compatibility of the elements.

The first stage 'Know the marketplace' is essential because the strategy adopted must fit into the market's expectations. Incomplete knowledge in this area will no doubt produce a strategy that will need considerable modification as the errors unfold in time. Given this information the next stage is to set numerical objectives.

It is important that all targets are measurable so the extent of achievement is communicated to all. This is the easiest stage and some companies have been known to stop at this stage and expect the objectives to be obtained without the planning and resource implications known. The other stages are used to prove overall feasibility, which is the essential outcome desired.

Many types of objectives/targets can be set; a few are listed below to give an insight of which can be selected. The numbers are chosen depending upon the present performance and the achievements of the competitors. Numbers quoted here are arbitrary.

- Increase Customer Number by 30 per cent.
- Introduce four new products per year.
- Reduce work in progress by 20 per cent.
- Improve stock turnovers to 12 per year.
- Sales per employee to rise to £50,000.
- Reduce scrap rates on nominated components by 50 per cent (usually the more expensive ones).
- Replace or refurbish ten machine tools per year.
- Reduce total area used by 20 per cent.
- Increase sales turnover by 10 per cent.

- Purchase 10 per cent more parts from supplies for the same money.
- Keep wage bill per person inside inflation percentage.

Once these **key** targets have been selected and set, which will position the company where it should be, the next stage can be entered.

Policies or plans now need to be formulated on how each objective is to be attained. For example, how can our customer base be increased by 30 per cent? If satisfactory answers cannot be found then the objective is unobtainable and has to be changed to a practical target. Hence the need for two-way arrows.

Each feasible plan or policy considered has implications for resources; they are initially evaluated at this time, but will be dealt with in more detail later. By these interactive stages a balance between required objectives and practical plans is obtained.

The next stage is to find the consequences of the plans on the business unit's resources of materials, manpower and machines. Some of the effects could be that extra skills are needed – more equipment, more shifts or more materials and space. Each of these has cost implications which will be considered shortly. It can happen that a specialised piece of equipment will be overloaded by the new pattern of production and alternatives such as invest in duplication, extra shifts, buy extra products from suppliers or alter design and processing methods used will all need careful consideration.

If insurmountable difficulties are confronted then the plans and/or objectives will again need changing to suit the circumstances revealed by this process.

Finally, the full cost implications are evaluated at the budgeting stage. Not only is the total cost important but also the schedule of when payments are made and returns are received. This financial commitment is then referred back to the Board of Directors as a business package proposal. The whole process may need to be reviewed if severe objections are made at this point.

Introduction to Problems of the Automobile Component Group (JHC Autoparts Ltd)

The group has been affected by the recession, particularly in the automobile original equipment manufacture (OEM) market. Also a policy of minimum investment in response to a declining market has left the group of companies with rising costs and low to non-existent profits resulting from the lower sales achieved. The results are not consistent, some parts of the group may be profitable, it is thought, but the accountancy and cost system does not help differentiation. These is a large indirect cost, centrally analysed, which is distributed over the products and hides the truth even from those wanting to effect cost savings from investment (see Chapter 8).

Refer to the organisation structure (Figure 1.1) to understand the wide range of activities involved and the need to rationalise the total group to prevent closure or even the receivers being put in charge of the company affairs from angry debtors' actions.

The list of OEM product business units includes horns, windscreen wiper motors, lighting units, alternators and starters, distributors, actuators, brakes and engine management systems.

The Board of Directors has come to terms with the fact that the business in its present form is finished and a radical rethink, to produce the strategy for the future, is called for. Many proposals are initially considered, ranging from sell off some business units or close down the complete group to expensive risk strategies based on a new customer base for car (OEM) sales being created.

Only one main strategy can be pursued to achieve a desired result. The JHC Autoparts Group Board must set the thinking in motion to achieve a feasible workable plan that maximises the full business potential, however large or small. This case gives later the steps that the Board and others took; would they be yours, will you agree with the policies adopted? Like all proposed changes, various opinions were strongly held but no one route can be known to be completely right; a need existed to have a strategy that was good enough to obtain a reasonable level of success.

MEMO
To: Reader From: Author
Strategy Formulation

A team of three consultants from the Industrial Services Department, leader Hugo DeWitt, has been chosen to carry out an in-depth study and make recommendations concerning alternative strategies available to the JHC Group given the circumstances they establish are evident. Consider yourself as part of this team and make a contribution to the suggested outcomes. Check your recommendations, when formulated, by comparing them against the decisions that were taken and carried out. Set questions at the end of the report may help to formulate your responses to the team.

Present Position Outlined – Historical Background

The JHC Autoparts Group business had peaked during the 1960s, providing a major share of the automobile OEM for the European Market along with other important outlets around the world. This success was built upon the pillars of rigid standardisation, volume production, up-to-date designs, reasonable quality levels along with a low pricing policy. Large profits from this period were used in other large business units of the Saltward Main Group to source innovation in a different sector.

This activity is sometimes likened to using a business as a 'cash cow', milking it while they can, but did they feed the cow enough?

Even though a market appears stable, some dynamic always exists and this was ignored to the company's peril. During the problems of the 1970s, which were many, inflation and over-manning had a great effect and competition increased, particularly on price which had been a main selling point. Competitors were using the investment in technology to make themselves more price and quality attractive. Less standardisation, more choice and product design changes also put a strain on the group's marketing of the product range. These pressures were not responded to, hence the lost customers. A new JHC Group Strategy to combat the competition was needed, but the initial response was only the same as before. This is sometimes called 'production orientation' rather than the more responsive 'market orientation'. The latter is based on ideas and questions, such as who are we making the components for? Ourselves or our customers?

Once this downward spiral was reinforced the results were a loss of the customer base, losing the contracts to provide the components on a newer range of cars because of investment policy and obsolete designs. The business refused to supply small quantities per week, when requested, because the group was geared up for mass production in a developing non-mass-production market. For example, different model styles of automobiles have different components assembled to them in the same body range.

Hence, the JHC Group arrived in the 1980s with many problems, which included losing money and having to be subsidised by the part of the group it helped to fund for innovation, old dedicated equipment in old buildings which were multi-level, creating long awkward travel routes for component movement, delays, queues, shortages of components for assembly, despite huge stocks of other parts, and quality problems linked to design, production and testing.

All these factors mean money is lost in delay, scrap and rework. Also the low investment policy was carried out to the point where middle management were completely demoralised by what was perceived as the indifference of higher management. This was shown particularly in regular rejections of proposed modifications to production which would save money in the long term. When top management gives up then eventually everyone else will!

Starting Point of Strategy Formulation

Hugo DeWitt and his team established their starting position and found all the above factors had brought the group to its present low position, which in 1986 included problems such as:

1. Outdated designs when measured against the competitors for cost and quality.

2. Unsuitable manufacturing methods based on volume production of standard products not designed for the new smaller specialised market.
3. Dedicated equipment to produce a steady stream of 'fixed design' parts, resulting in poor flexibility and response to customer needs.
4. Outdated multi-level buildings with low ceilings, poor lighting, queues at lifts and general transport difficulties.
5. Many different sites, some providing parts for other sites.
6. Large stocks and work in progress which were far higher than those of competitors.
7. Poor sales per employee ratio compared with competitors.
8. Warranty claims twice as high as competitors.
9. Relatively few customers, some of which were in financial difficulties.
10. Verification of quality effectiveness not yet attained but demanded by customers.
11. Intense competition on certain components. Decisions required based on whether to keep producing the whole range or to 'buy out' parts from previous competitors.
12. Some of the products were based on easily duplicated low technological content with no developments expected for many years.
13. A large pyramid management structure comprising many levels, at least nine, with unclear decision-making responsibility. Who controlled the budgets was not clear to some levels of management (communication problems)
14. No one seemed specifically to own any of the above problems because of centralisation of management control.

The overall problem to which a strategy must be addressed

The main problem facing the JHC Group Board of Directors was how to proceed in the light of incomplete detailed information, facing the wide-ranging nightmare, which had existed for many years, with limited information suitable for analysis because of the central accountancy system. This system was built upon the idea of *group overheads* which treated all parts in a similar manner in terms of cost, that is using just one distribution method. The 'real money' spent on components as they travel around this complex system was distorted by the blanket allocations of indirect costs/overheads. The bottom line figures of the business revealed the truth but an explanation of each product's individual contribution or loss was missing, or when available, of doubtful origin.

The idea of collecting viable information from a modified costing system was considered to be too slow to be effective in the circumstances and time availability. A more dynamic 'quick and dirty' approach was needed. Hence a radical 'decision criterion' that could quickly be applied to make things happen was required.

E. Questions for Reader and References

This is clearly an open-ended problem and no one would expect different people with this limited information to arrive at exactly the same strategy. However, consideration of alternatives is the basis of all strategy formulation (just like chess) and should be applied here. This is part of the systematic approach adopted.

Consider the possibilities available

1. How many different basic alternatives can be listed for the Group Board to consider?
2. How should the Board proceed to evaluate the different alternatives?
3. List any criteria that the strategy formulation can or should encompass.
4. Outline a strategic plan that appears feasible in the circumstances noted.

Commit the ideas considered to paper so they are reasonably structured. Experience shows that random thoughts, with no record made, tend to be self-serving and uncritical of performance.

Suggested references for further information

Competitive Advantage of Nations, M.E. Porter, Macmillan (1990).
Concepts of Strategic Management, L.L. Byars, 3rd edn, Addison-Wesley Longman (1991).
Operations Management – Concepts, Methods and Strategies, M.A. Vonderembse and G.P. White, 2nd edn, West, USA (1990).
Strategic Management – Concepts and Cases, J.A. Barnett, Wadsworth International Thomson (1988).
Strategic Management for Decision Making, M.J. Stahl and D.W. Grigsby, PWS International Thomson (1992).

F. Proposals Adopted by the Group Board

The team led by Hugo DeWitt made recommendations which the Board immediately adopted. Not surprisingly, the Board chose to change using its present policies and to pursue a new strategic planning direction. This was a complete U-turn which demonstrated the severity of the problems faced and showed a new commitment to survival action.

An outline of the new strategy set by the consultants can be listed as follows:

- Break up the group, with each business unit having to stand or fall on its own merits.

- Raise money to fund the changes in the remaining product areas by selling firms making non-strategic products.
- The criterion on which decisions about what business units shoud be sold off to be 'technological development' potential. Products that are technologically based are considered better earners of profits because they have less competition and higher value added returns. Too many low technology products suffer severe competition from low-labour-cost/subsidised companies from various parts of the world.
- Business Units to be sold by the above criterion to include horns, windscreen wipers, motors, lighting and alternators; each of which has reached the end of its main technological development.
- Products to be considered for further analysis to include distributors, sensors, actuators, braking and engine management systems. These were considered to be subject to reasonable technological change and sales potential, and would deliver high added value in the marketplace.
- A task force to be set up for each of the above chosen product groups and instructed to evaluate individual business potential which would ideally be situated on separate single-level sites.
- Vacate the main factory/multi-level units because of their unsuitability for world-class manufacture image.
- Each Business Unit to be autonomous and obtain sales for itself, and hence deliver a profit for the main group. If it could not justify this achievement, it was to be disposed of.
- Each task force to consist of about six people from different disciplines. This would give wider expertise to enable a comprehensive business plant to be produced.
- Each task force to make a bid, based on its business plan, for the monies available from the sales of the other Business Units.
- The better the bid, the more comprehensive the plan, and the more certainty of acceptance by the Board. However, once given the go-ahead, the Group Board would monitor achievements versus the plan and take action where necessary.

The thinking behind these strategic manoeuvres was to look to the future and not bother to analyse the present problems in detail. This stage represented a new way of operation for middle management. The old centralised decision-making process was replaced by local management structures which have ownership over the decisions. The aim was to escape from the old multi-level premises and group overheads and change to single-site operations which would clarify the inputs and outputs of cash flow, while at the same time containing costs to the area that created them.

Each business plan was for five years and included estimates for sales, expenditures, new products, new customers and planned profit from the expected advances in methods and technology. Evidence for these claims had to be found and substantiated. This was the keystone of how the group would obtain feasible strategies for feasible business units.

An Example of Outcomes that One Task Force Accomplished in a Business Unit

The following business plan was accepted by the JHC Group Board and gave the following scenario for one of the main OEM producers, Duncans Engine Controllers. The information that is given is for the particular business unit and used to support the hierarchy of strategies (see Figure 2.1). It can be seen that the mission statement quoted requires a number of levels of support. The details given from this business unit case study are used to support the overall corporate strategy and hence ensure, in part, that success is achieved. These particular strategies and achievements are expounded for the reader's interest and to allow full consideration of the multi-levels of strategic planning and control, even though outside the basic questions set earlier.

To improve the Sales/Employee ratio, certain plans were made because this was seen as a critical measure of company performance. These involve the reduction of the in-house workforce from 652 to 438 after leaving the old multi-level site two miles away. This labour reduction was accomplished by interviewing each prospective worker for either a job or offering good redundancy terms. Complementing this policy a further outlet for labour was employed, namely one of providing both machines and workers to sub-contractors who took over some of the previous product range. Hence 'make' became 'buy', which simplified the tasks at hand.

It was considered important that the morale of the selected workforce was high when they moved to the new site. The dangers of an alienated workforce as a result of the dramatic changes planned had to be addressed. Hence, time allowed for communication and listening were important ingredients. The interviews used were particularly to determine the willingness of individuals to work in a new multi-skill environment free from any form of demarcation. A flexible working pattern would best suit the new customers' needs as the company changed from dedicated to flexible outputs of small to large orders.

The marketing plan covered the expected demise of outdated products over the five years and highlighted the volumes expected from the replacement products.

Efficiency was to be a major factor; the thrust was to be for better equipment, a trained workforce, new methods and procedures.

All dedicated equipment was evaluated for suitability in the new flexible environment. An example of this was a particular Machining Centre, tooled up to produce just one part in thousands every week, which was unreliable and near the end of its economical life. A number of alternatives were considered:

1. replace the machine, cost £300,000;
2. refurbish the machine, cost £125,000;
3. replace with six CNC lathes £240,000;
4. other suggestions, which were not actively pursued.

Option 3 was eventually decided upon because it reinforced the strategy of flexibility and could tackle existing jobs as well as having an ability to produce the future product range, which was still being developed (see Chapter 8 for an investment case study).

The initial bid Duncans made for capital requirements to the JHC Group Board for the projected move to the new site and for equipment replacement was in the order of £2.25 million. However, only £1.6 million was made available from the Board and the task force was asked to modify the plan to make it feasible at this figure. This was eventually achieved by changing the basis of machine improvement. Each of the hundred most important machines was rigorously tested for process capability using statistical methods. A wide range of capabilities was discovered over the range. These tests enabled decisions regarding replacement or refurbishment to be made; the rest of the equipment could wait for another year's capital expenditure budget. Refurbishment was extensively used even though it caused extra problems; these included keeping production figures normal while important machines were absent for one to three months. The refurbished machines were not accepted back until they had passed rigorous process capability tests which were included in their required quality specification sent to the subcontractors.

The methods and procedures adopted to improve efficiency included the important move of dividing the single-level floor into four main units;

(a) Supply
(b) Machining
(c) Subassembly
(d) Final Assembly and Test.

Each is the supplier or customer of the adjacent unit, so this established ownership and responsibility for each unit controller. These managers were given wide-ranging power inside their areas, which was totally different from the old regime. Japanese methods of production control were planned into the system, such as Just in Time, Kanban and Visual Controls, to aid flexibility, reduce work in progress and increase stock turnovers per year to twenty four. See other case studies for further details (Chapters 6 and 7).

Outcomes

New product ranges attracted new customers to Duncans, as did the flexible manufacturing policy which turned the old 'high volume dedication' policy on its head and created an extra range of low volume products which could be flexibly and efficiently produced. No order was turned away, a quote and delivery date being given to all potential customers: not revolutionary elsewhere but new in this business unit.

It took one year from setting up the task force to full production output in a new rented factory built on a greenfield site. This was achieved according to plan, timescale and cost forecasts. The overall results achieved can be summarised by the following list of objectives which supported the corporate strategy:

- Sales per employee up 35 per cent.
- Six new products launched in prototype stage.
- Stock turnovers doubled from 12 to 24.
- Scrap and rework reduced by 75 per cent.
- Warranty claims halved, in line with competitors' performance.
- Fourteen unions in the 'old buildings' now reduced to three on new site. Negotiations easier and faster.
- Three grades of worker established, compared with seventeen in 'old building'. The three grades were based on the level of skill flexibility. All workers can be on the top grade, with no artificial limits set by cash limits.
- One canteen for all, instead of the four previously used.
- The design office is now dedicated to one business unit and does not provide for other departments' product range. Better response obtained. Lead-time halved.
- Overhead costs reduced dramatically along with profit targets achieved.
- Three levels of management, each with ownership; clearly identifies responsibility, helping to remove causes of problems, not the symptoms as before.
- Verification by five different quality standard authorities achieved.

Summary of Reasons Learned from Problems Analysis

The problems of the past were considerable and radical action was needed for survival. The first significant step was the initial sorting of the potential cash growth 'rising stars' products from the 'dogs' (obsolete and low profit ones). This was achieved by selling off poor performers to provide resources for the others. The second step concerned the present moribund system of costing – it was swept aside by the use of the new strategy, built on future prospects alone and making the business units autonomous particularly with regard to their costs. The middle management morale problem was removed by giving it the clear responsibility of producing plans for success or, if that were not possible, then closure.

Starting on a new site with a 'new' workforce trained and mobile with updated equipment, as Duncans did, produces a good combination of assets to face the challenges of autonomy. Particularly of note was the time and money used to make the personnel changes smooth and without too much rancour.

In fact, each one of these objectives achieved is like a 'brick' that enables the rest of the wall to stand. Each department has a contribution to make and these build up to make the total changes that enable the strategy 'gap' to be closed.

The attention to detail and identification of the essential factors of the business success are in stark contrast to the original long-standing strategies adopted by top management.

Another scenario would require another strategy and, as each scenario changes, so must the strategy. Finally, needless to say, the surviving business units are using continuous improvement schemes to satisfy and fine-tune the new operating conditions they face. Change is an accepted part of all the business unit's operation and future success.

Text Reference

1. Thomas J. Peters and Robert H. Waterman, Jr., *In Search of Excellence*, Harper & Row, New York (1982).

3 Organisation Structure Design

Section

A. Learning Outcomes

The importance of organisation structure and design is often ignored or not discussed by people who do not fully understand what is achieved by the appropriate use of systems.

This case study will demonstrate that:

1. When alternative organisation designs are considered, the choice made will affect the top executives down to the shopfloor.
2. The structure is used for decision making and communicating these to others.
3. Though it cannot be seen or touched, its existence dominates a firm's operation.

4. No single system can answer all companies' needs because it is a complex unique activity, often Chief Executive based.
5. As objectives and strategy change, so should the organisation used to control the required outcomes.

B. Learning Objectives

After studying and analysing these case studies the reader should be able to:

1. Explain the importance of organisation structure from strategy to potential success.
2. Identify the key features of an organisation chart.
3. Define the different basic structures on which each design is made.
4. Understand the effect of the organisation design on functions and individuals in a company.

C. The Basics of Organisation Structure

Introduction

All businesses produce and market their products and services through an organisational structure. This structure can either help or hinder a business in achieving its objectives. It has been said: 'management is about being responsible for more work than can be accomplished by one person'. This statement highlights that, to complete the manager's responsibility, the work must be done by others through an appropriate system. The essential components of good management include delegation, control, organisation design and accountability. An essential contribution comes via the organisation structure which is designed, just like a machine, to fulfil particular needs. Many interlocking management functions will come together in this structure and hopefully the whole will be stronger than its parts. It is extremely complex in many instances and only partial generic theories apply. In practice every organisation structure is essentially unique. The specialised nature comes from the combination of marketing objectives, products offered and company size creating a different design. When the extra dimension of people and their attitudes are added, the potential for problems through incorrect operation is established.

These intangible networks are waiting to be discovered and understood by all newcomers to the company who want to make their mark. The ability to understand how it works plus how to use informal shortcuts through its hierarchy are the key to actions to get things done. Without a reasonable knowledge of the prevailing systems the chances of success are low.

Conversely, many people dislike organisation theory and design, considering it 'boring' and irrelevant. Nothing could be further from the truth. It can be the managers strongest weapon for making changes and can form an immovable resistance when generally ignored. Every management newcomer to a situation needs to ask many questions and use all the recorded information available to employ correct procedures to prevent resistance. Many companies document the procedures in manuals but the reluctance to open them and use them is considerable. Some people think one of the reasons organisations do not operate successfully is because of the lack of insight about how the designed system works. It is not good to have a poor system, but also a good system can be destroyed by the apathy of some. What is the way then to success through organisations? One theory adopted over the past few years is based on the idea that responsibility for a system and its success should be 'owned' by one person in each area of the company. This results in the propagation of smaller departments and production cells under the control of leaders with specific responsibilities, who work in the area of their control. Often they are referred to as *cell leaders* which means they 'head' the team. This is a step towards decentralisation on the shopfloor.

The need for a designed structure is therefore based on higher management's requirements which enable their policies and plans to be carried out with a minimum of waste and delay. Whatever some critics may say about 'red tape' and bureaucracy, in the final analysis, without the organisation there is only chaos and waste. It is up to those involved to ensure that guidelines and structure combine to provide the basis for the passage of information up and down the organisation. Only this will ensure that the products or service offered meet the customers' needs.

One way of showing the interaction and components of this transmission of information is given in Figure 3.1 (on page 29). The components need to be balanced so that an overall effective result is obtained. Anything other than a balance is unstable and is bound to fail.

Types of Structure Used

Many books have been written on this subject and the following is a personal outline summary.

Organisation structures are designed to fulfil specific objectives and are often based on a particular theme. This chosen theme may be one of the following:

(a) Functions of the company,
(b) Products/services of the company,
(c) Matrix organisation,
(d) Geography of company activities – a mixture of any of the above.

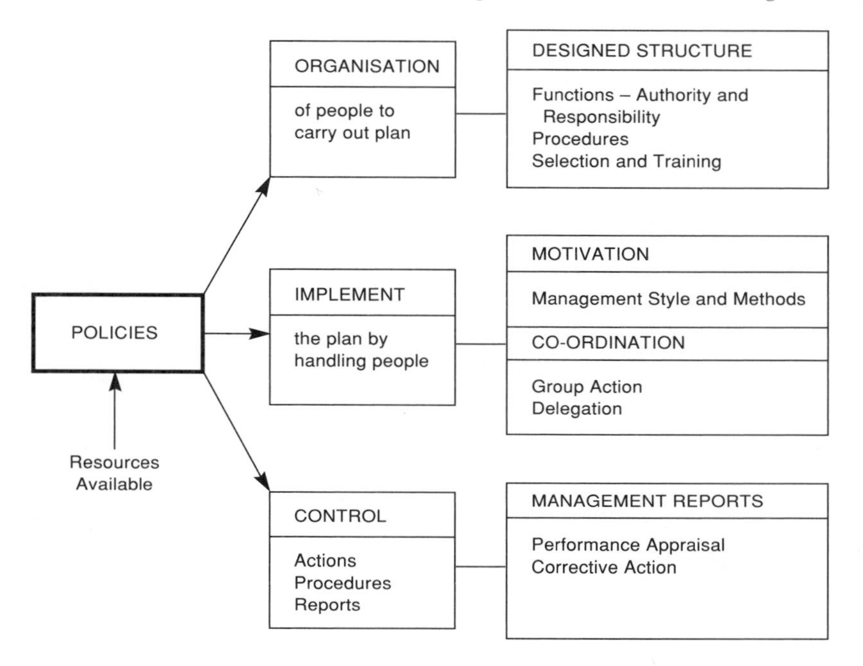

Figure 3.1 Using the organisation to carry out the policies of an agreed strategy

(a) Functional structure

Companies that manufacture components have a raft of functions they need to complete for each product, no matter what the product diversification. These include Sales, Manufacture, Accounts, Personnel, Engineering, Design, Maintenance etc. An organisation designed around these functions, see Figures 3.2a and 3.2b, will possess strengths which will concentrate all knowledge in one area, and people with similar job responsibilities can be in close contact with each other. Some functions have general headings with specific functions shown lower in the organisation chart. The components of organisation, style and feedback reporting enable the plans and policies to achieve the company objectives. This represents success when each one is right for the set constraints which apply.

(b) Product structure

Larger companies can have a wide range of products and some may be totally dissimilar; the product types can then be the design parameter theme for the organisation. The products would be autonomous and self-organising. Consequently, functions would be duplicated in each of the divisions, for example Product A and Product B could both need Sales, Design,

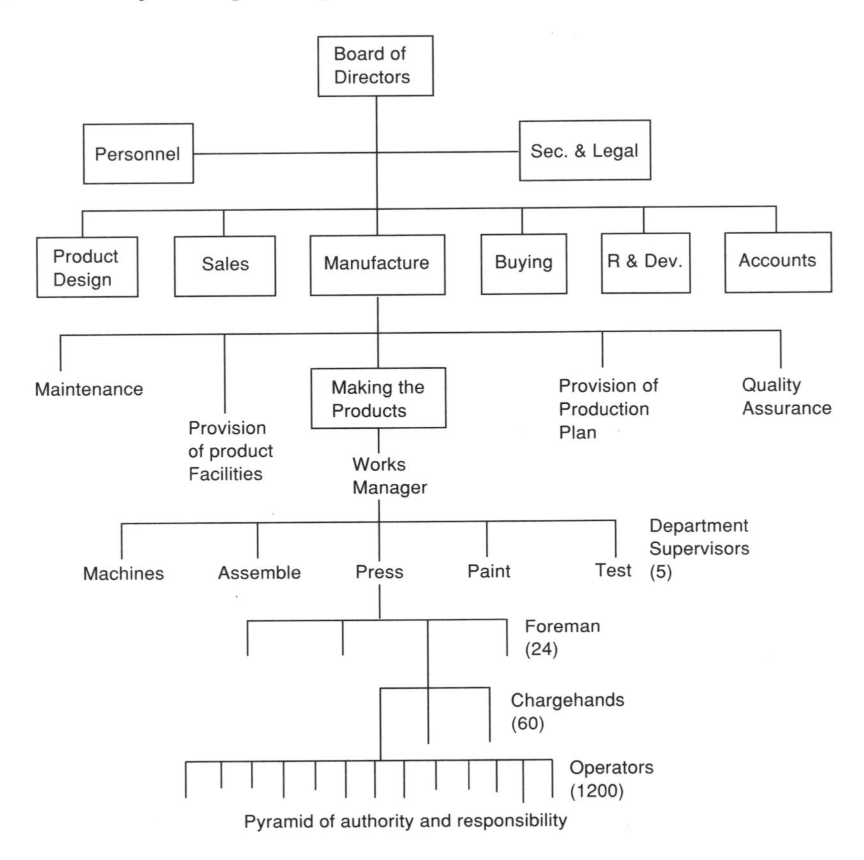

Figure 3.2a A company organisation based on functions

Manufacture etc., so in this case the Design function of the company is split into each product area. The concentration of staff on one set of products is good but it is at the price of duplication, see Figure 3.3.

Duplication is a common feature in many organisation designs, and is the reason why the objectives of management need to be clearly established. For example, Sales can be a problem if one customer orders from more than one division. To try to get the benefits of both, some companies adopt a matrix system (Figure 3.4), in which function and product are jointly 'owned' by higher management.

In an attempt to gain the benefits of both systems, a link between these two diverse areas is created. A manager is appointed to oversee the operation of all functions and product responsibilities in one area. Clashes for resources can occur but the emphasis is on integration and teamwork.

One way of showing more localised information for the worker concerned in a particular part of the organisational structure is demonstrated

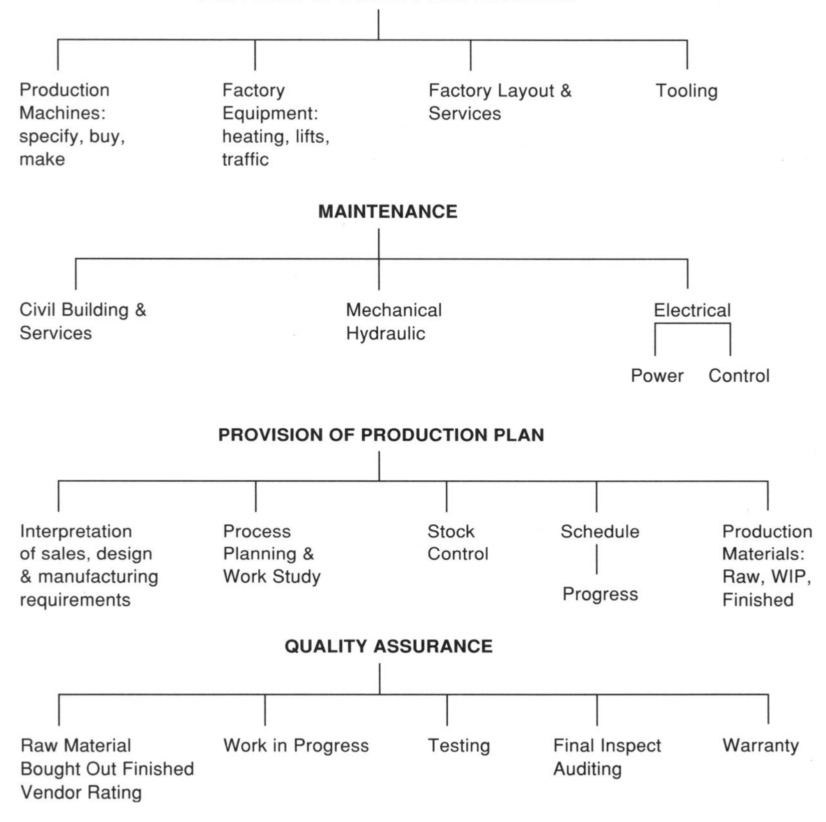

Figure 3.2b Activities of some of the organisation functions

in Figure 3.5. Here the normal line relationship is shown but in addition the interactions with other areas that normally constitute working activity are identified. This bias towards particular departments assists new employees to understand where the expected communication channels are. The figure illustrated was used to induct a new recruit into an organisation and had beneficial results.

Another feature of organisational design that needs attention concerns how many layers of management are required. Layers of management also decide the length of the management 'communication chain'. Circumstances, after analysis, should decide how many managers are required for optimum organisational performance. The effects of different design decisions are demonstrated in Figure 3.6. In this figure, a thousand workers are shown in two types of organisation, each of which has its advantages but also creates totally different organisation cultures.

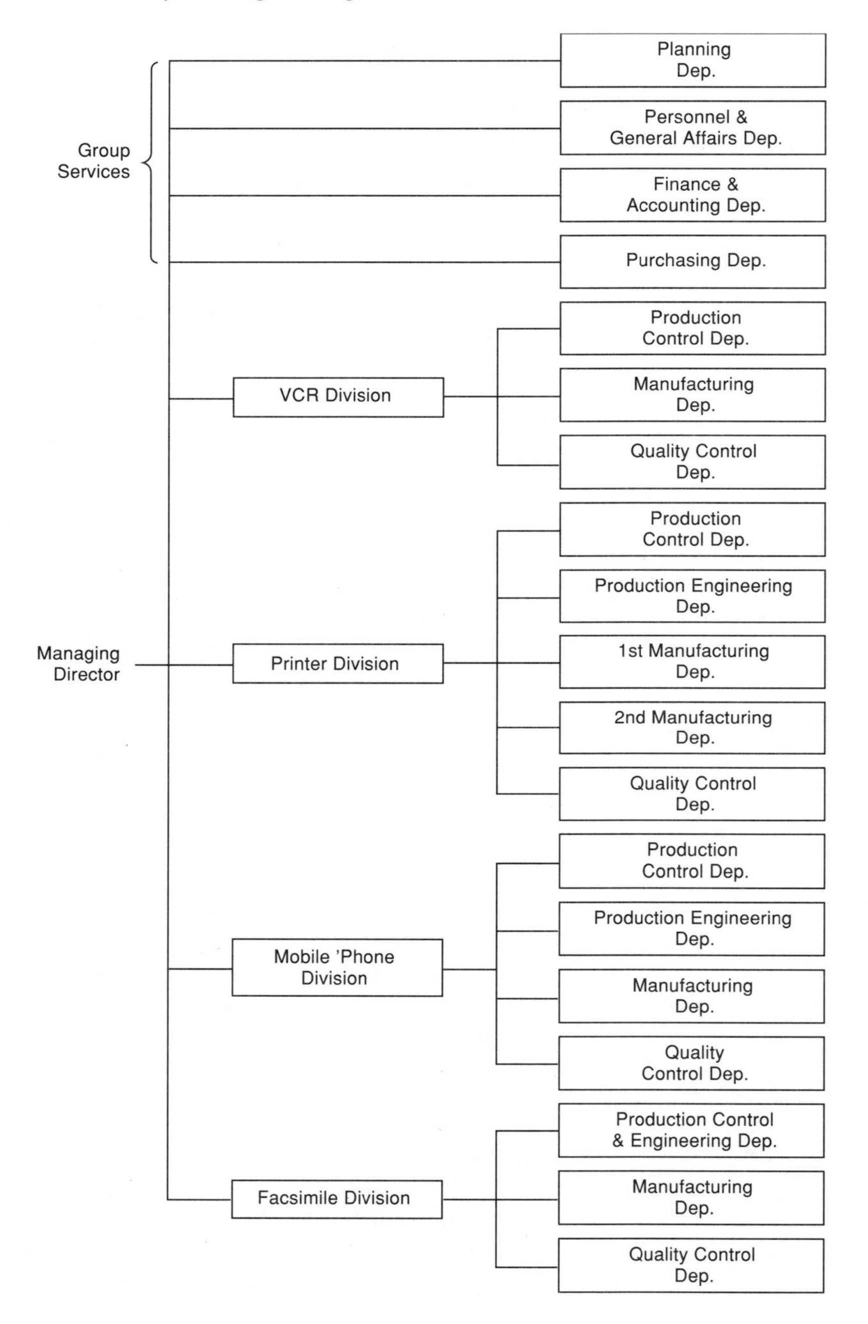

Figure 3.3 Product-based organisation chart

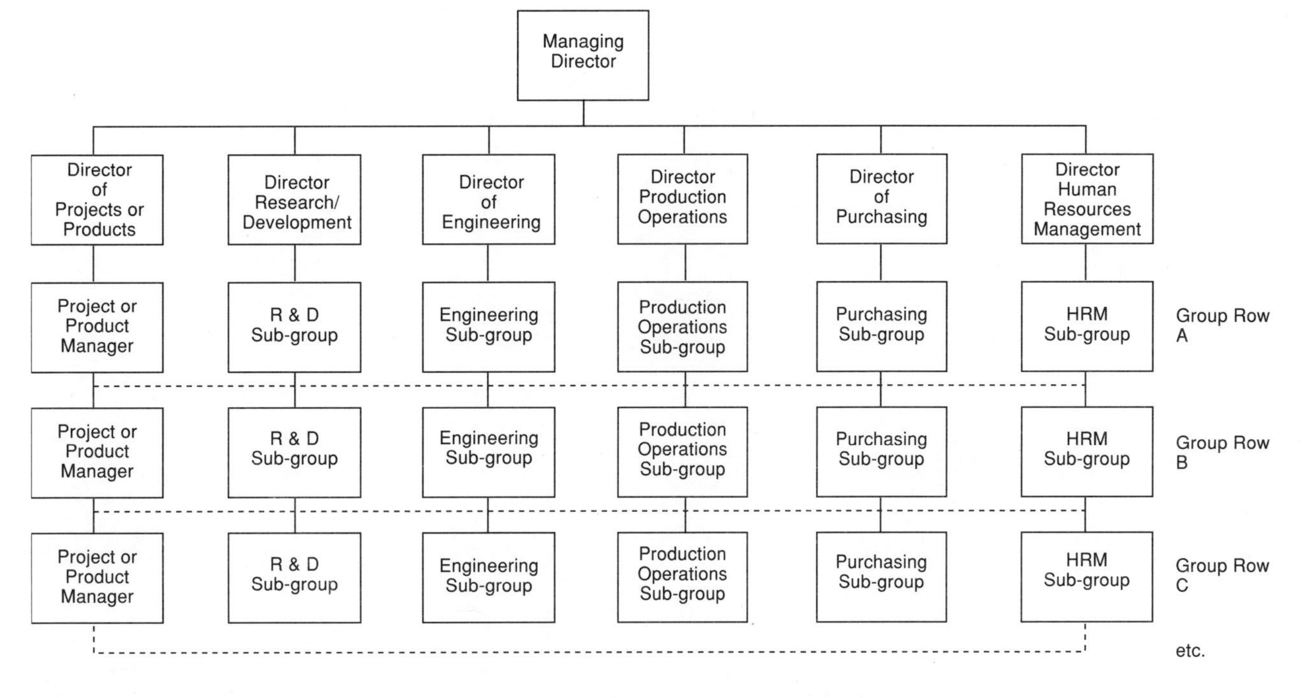

Figure 3.4 Matrix departmental structure for project- or product-based organisation

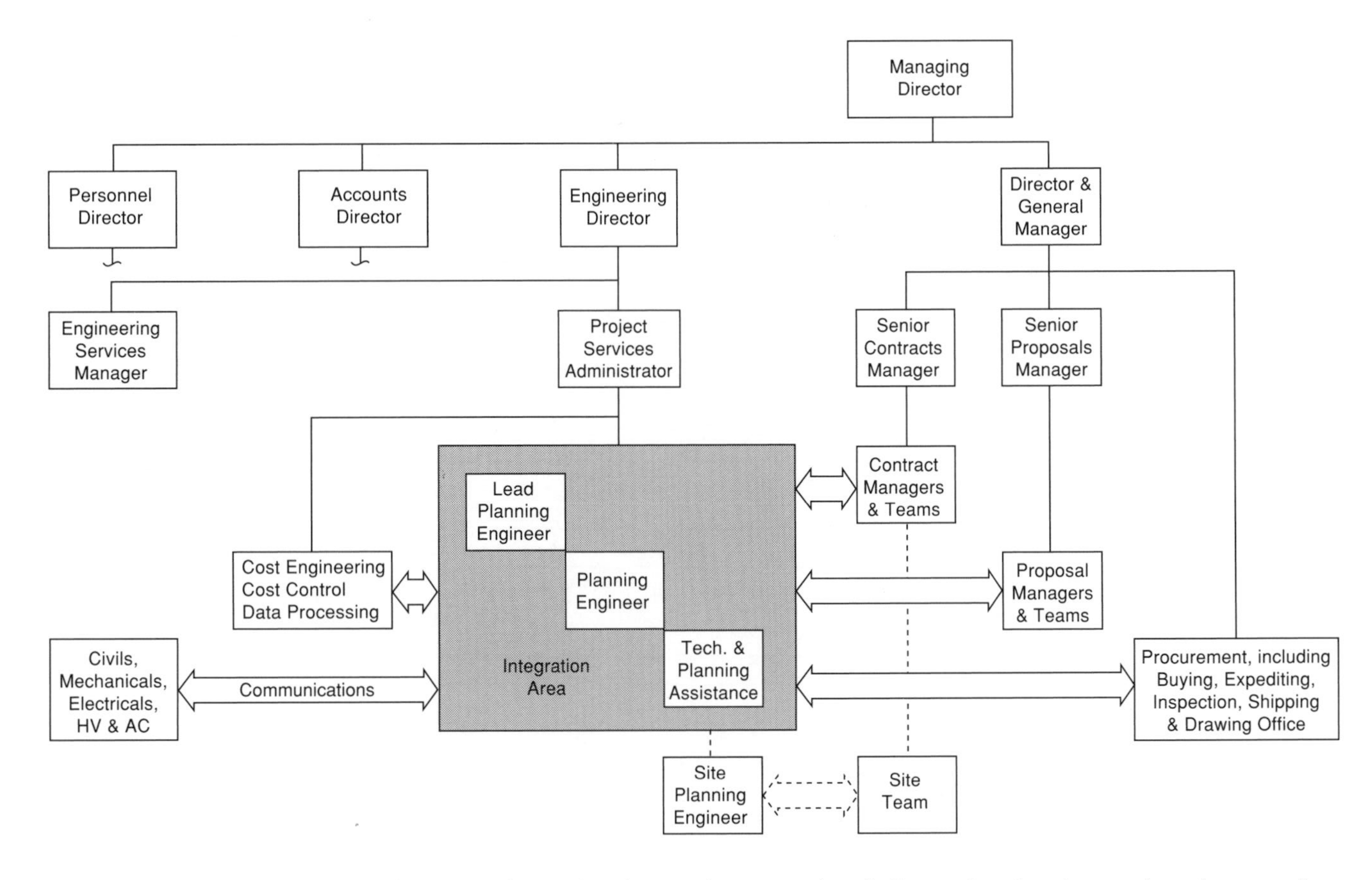

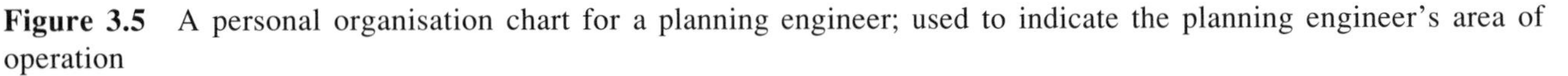

Figure 3.5 A personal organisation chart for a planning engineer; used to indicate the planning engineer's area of operation

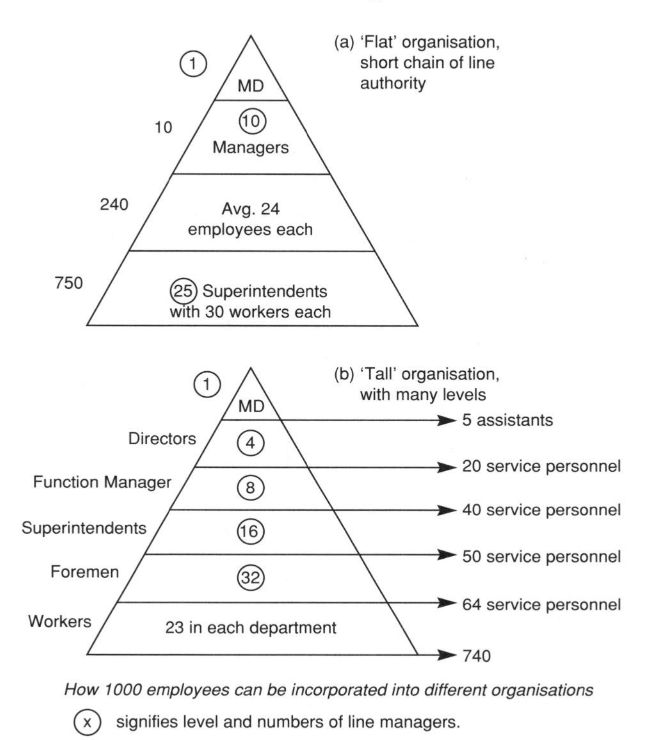

Figure 3.6 (a) 'Flat' and (b) 'tall' approaches to organisational design

The number of people reporting to each manager is different, so this affects the 'distance' at which decisions are taken from the identified need. This ranges from one level of management to several, resulting in decision making far away from the situation.

It is not unknown for an engineer to request £25,000 for an investment in equipment, only to discover that the final decision of 'yes' or 'no' was taken five levels of management higher in the organisation. The actual quality of communication between the user and the decision makers is often, not surprisingly, low. For instance, in the normal operation of a company, manager 'D' sees his boss, manager 'C', regularly in day-to-day operation. Occasionally manager 'D' may see manager 'B' with manager 'C' in connection with work problems. The chance of manager 'D' discussing anything with manager 'A' is remote because manager 'B' would be accompanied by manager 'C' instead. Anything else would marginalise managers 'B' and 'C'. An alternative is to arrange meetings of different layers of management in which decisions are taken. This only comes about when manager 'A' has this particular style of management: a rare event. Hence, built-in short horizons are common in organisational design. They are only expanded if the company deliberately uses 'vertical' management meetings to discuss problems, rather than the more usual 'horizontal' ones.

Unfortunately, managers, knowing their place in the organisation, fixed by their position on the organisation chart, can easily become departmentalised in problem-solving activity. Upper management often responds by trying new ways to get round this mindblock. One of the methods used recently is Concurrent/Simultaneous Engineering which uses matrix teamwork across departmental boundaries to shorten the length of decision making. Time does not allow a full analysis; use the references to help if required.

Decision making is a critical feature of the designed structure. In the case of only a few levels of management being used, the question and the answer come face to face and acceptance of this is easier to sell to those involved. However, the multi-level company suffers by separating the question and answer, and risk causing discontent through lack of information at both ends of a long chain of activity.

Ownership of functions and hence their problems have been shown to overcome the decision chain because the responsibility has been delegated and clearly identified.

Centralised/Decentralised Options

Responsibility and Authority are constantly evolving in the organisation, from either deliberate changes or from a manager's stronger personality/attributes. Whether to use centralisation or decentralisation options can sound dramatically different but often the option only relates to one level of management, that is to one more or one less. It is often simplified to either delegation down a level or taken back one level. Centralisation is upward movement and decentralisation downward movement of responsibility. Nevertheless, it is no less significant for this one level of change. Moving the responsibility for a budget down one level alters the fundamental thinking of the people involved. The lower manager now has the need to plan and control resources while the upper one is released from day-to-day activities to consider the future direction and strategy of the section. To the outsider, or even the examiner of an organisation chart, the changes may seem small (one level) but organisational knowledge and analysis will reveal more. When centralisation of one level occurs it is just as dramatic. Responsibility for a budget is removed to a higher authority and it is as if everyone has been marginalised and demoted to doing as they are now told.

Designed for a Purpose

Just a few of the multi-faceted aspects of organisation structure have been looked at. For some operational information it is useful to view the business pages of any quality newspaper. Articles show changes in personnel and restructuring of groups of companies all the time. The dynamic nature of structure, even if initially only at the top, will eventually affect everyone in some way. A typical example is when a Chief Executive leaves, for whatever reason – retirement, resignation, being sacked or promoted elsewhere. The replacement has two choices, to run the business to the same pattern as before even though he wants to be his 'own man', or to change and develop his own agenda, objectives and strategy to *improve* the company performance. To achieve this desired outcome it is necessary to change personnel, alter structure and either expand or down size. The structure will be an essential part of his plans and this is when centralisation/decentralisation is seen, in practice perhaps with either function to product based or matrix hybrids being developed.

An investigation into the history of local companies will no doubt reveal these type of changes. Changes are not restricted to when Chief Executives leave; many incumbents have needed to tackle all of these problems to ensure the survival of the company. As the need changes so does the structure. Any company that has not redesigned its structure for years must be operating under a totally different set of market forces to those which most companies endure. The management style of the Chief Executive is a strong feature of organisation design. Even the choice of managers for positions at different levels influences the operations. If the choice is between, say, people with attributes of meticulous researcher, people pleaser, committee manipulator and backside kicker (abrasive), then the outcome tells much about the upper management's requirements.

As no one has found the perfect organisation structure, everyone is looking for improvement in what they have. New ideas are tried continuously to achieve a breakthrough in efficiency. These new ideas can be almost radical because most of the history of modern large organisations is only 100 years old. It is still developing at a pace because the demands have changed so significantly over this time.

Before the changes in design can be formulated it is necessary to collect factual information so that only realistic design is considered. One of the better ways of finding this out is to use the technique of auditing.

Organisation Audit

Financial audits and quality audits are in common practice and reveal problems of malfunction or verify that all is well in various sections. In the same way an organisation audit can be just as revealing about adoption of strategy or knowledge of objectives held at various levels.

In theory all is well: the Board of Directors approves the strategy; the Managing Director and his team consider how it will be implemented; then through the organisation each member of the team, working as a functional head, determines how the department will do its part. However, some audits have revealed the following facts about companies who do not follow the theory correctly.

Many managements:

1. have not thought out what needs to be done if targets are to be achieved;
2. have no real idea of the many problems with which many of the employers are struggling;
3. have little idea of what the problems they are 'living with' are costing;
4. do not relate production and other problems to business results and are not therefore prepared to make the necessary investment in solving problems;
5. are more concerned with firefighting than fire prevention;
6. have come to live with high costs and low productivity by waiting for problems to occur and then buying their way out;
7. are always surprised when a new difficulty or unexpected (though predictable) problem occurs.

Hopefully, not all of these statements are true of any one company; they demonstrate the audit's potential to provide the facts of organisational operation. Just like other audits they rely on questioning that becomes more probing and detailed in the light of answers given.

The actual questions used obviously vary from one organisation level to another and also from function to function. They are often based on the following framework. First comes an initial exploration of targets; what targets have been set, how they relate to company strategy and what thoughts have been given to achieving them.

Here are some examples of checklist questions which may be used:

(a) What are you trying to achieve this year that in any way differs from what was done last year?
(b) What does this mean in terms of increases or decreases from or within the functions or various departments?
(c) What difficulties do you foresee that might prevent the section from achieving the required output levels or quality?
(d) Are there any particular difficulties that remain only partially solved, carried over from last year?
(e) What would be the costs or what would happen if all these problems were not solved now?
(f) Is there anything that needs to be done at your level to facilitate overcoming these difficulties?
(g) Are there any major problems or decisions that your group meetings have to handle that may give some difficulty?
(h) Does this group as a whole or individually require additional skills, knowledge or understanding to enable it to handle its responsibilities?

(i) To achieve the targets, are there any problems that arose last year or new problems at present that will need to be overcome further down the line management?
(j) If you can overcome all the difficulties that have now been identified, are you absolutely confident that all targets will be achieved?
(k) If not, what other problems, difficulties or risks might get in the way of success?
(l) Are there any changes that have to be made? What are they, and where will they have to be designed and implemented?

A similar questioning technique to obtain facts is outlined in Chapter 9, where different information was needed.

D. Introduction to Problems at Edward Marks Ltd

A new Managing Director had recently taken over the firm and wanted to change the organisation structure he inherited from the previous Managing Director. The company performance had not been as good as expected, and potential outcomes had not materialised. The common reason given by a number of managers was poor communication. Detailed information and decisions were not fully understood and it seemed as if each department was pursuing its own internal targets.

To counter these failures an audit of the various levels of management was to be conducted. The new Managing Director thought it essential that a 'neutral' conducted this questioning process to prevent ingrained attitudes to influence the outcomes.

It was decided to contact Management Services at HQ and ask it to provide a person who could effectively collect data, analyse it and report back to the Managing Director with suggestions and alternatives. The Managing Director had his own ideas but thought it was politically sound to incorporate them into someone else's suggestions after a high-level investigation.

It was agreed that Susan Luong from Management Services would spend two weeks carrying out an organisation audit and then report two weeks later. In his short time at the company the Managing Director had noted symptoms that made him uneasy about the organisation design and operation. These included:

(a) a flat structure based upon 20 departmental heads, some being called director although no particular seniority was established;
(b) quality control and stores operations were split under different section heads;
(c) power struggles between managers of different departments because individual responsibility and ownerships were not defined in any of the job descriptions;

(d) some section heads made hasty decisions for others without informing those affected;
(e) at least five directors were known to be communicating freely with the same customers, thereby isolating the marketing section;
(f) decision making could be long and tedious with regular meetings involving too many people;
(g) even when decisions were made at meetings they could be interpreted in different ways because the common need was not identified;
(h) 'departmentalisation' was widespread, an attitude that promotes a narrow viewpoint which is not company based;
(i) innovation in new products was slowed down because no one had been allocated this specific responsibility. Lack of specific responsibility was also repeated in other areas.

The Present Position Established

When Susan arrived she formed a team consisting of the Marketing Director, the Production and Operations Managers, the Director of Special Products and Nathan Peters, a recently employed graduate in systems design.

Susan had brought with her a checklist of all the questions she wanted answering to fulfil her objectives. However, second opinions from the others would help, together with the sharing of the work load.

The set objectives of the team were clear, namely that the audit must:

1. Check that targets have been set within the framework of a strategic plan.
2. Have targets that are feasible and agreed (if not, the auditing team will get the targets set).
3. Identify the key problems and difficulties that need solving if targets are to be achieved.
4. Identify the cost of waste if the company fails to solve such problems.
5. Identify key changes required in organisation, systems used, working methods, skills and attitudes.
6. Identify the resources, assistance, training and development of staff concerned, in order to deal with each problem.

The essential concept behind the audit is that of changed circumstances. If the company expects and wants to perform exactly as it did last year, then this is not a problem and an audit is largely academic. However, if the company intends or expects to have to make changes, the problems need to be clear. The audit contains two main elements: what happened last year and what the future holds. Answers are required for:

(a) What do you want to do this year that will be different from last?
(b) What external influences or difficulties might make you change what you are currently doing?

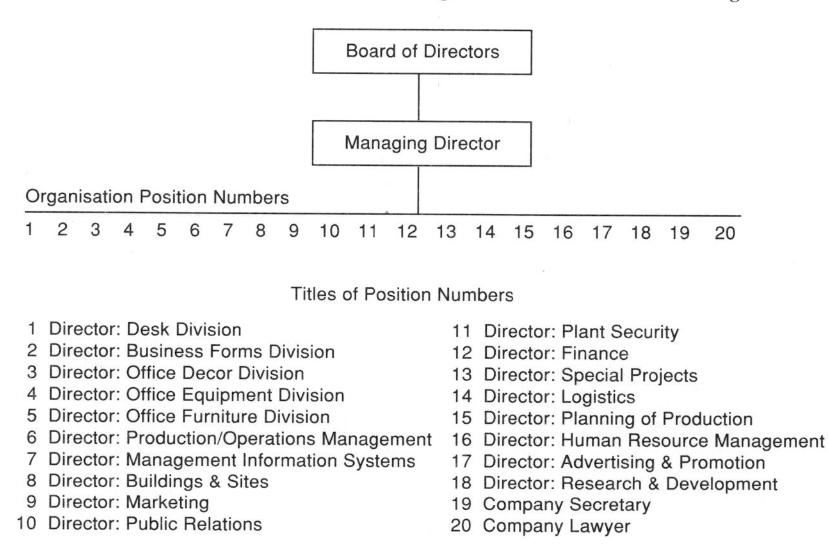

Figure 3.7 Example of a 'flat' organisation

Category (a) will also cover the difficulties encountered last year but not yet resolved. Items such as how to increase productivity, reduce breakdowns, improve industrial relations, eliminate quality problems, improve cash flow, optimise debt collection, reduce lead time for new products and set more realistic delivery dates would also be included.

The second category requires a systematic look into the future in order to anticipate such factors as competition status, market trends, market share, material prices (often involving currency fluctuations), changes in legislation, for example on environmental grounds, plus the degree of internal resistance to change offered by workforce and management. A thorough audit is thus a substantial exercise where problems are regularly found. It must start with the Managing Director and will eventually include heads of all major departments or functions. How far down the organisation it progresses depends on a number of factors.

Typical Problems Identified

The audit was so comprehensive and the analysis so wide that only a few examples of the questions raised are included to demonstrate the activity.

Situation 1

As the Managing Director thought, the flat nature of the organisation came under scrutiny and a simplified outline of the problem is given in Figure 3.7. Many alternatives are available to make this design taller. Consider this problem and then answer the questions set later.

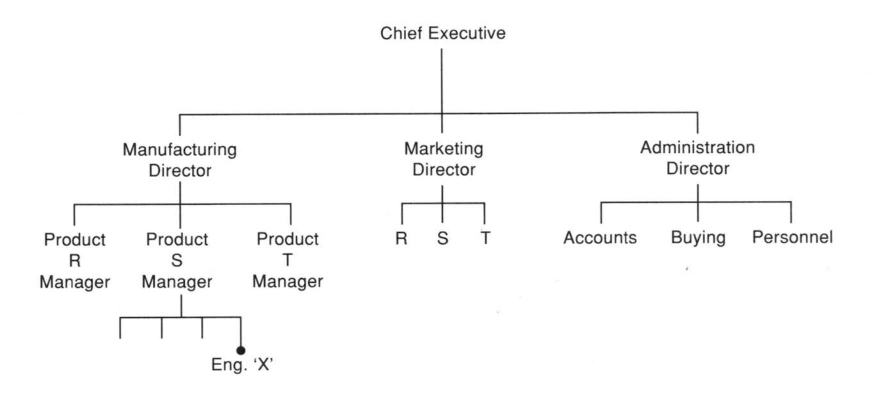

Figure 3.8a Situation 2: function-based 'A' structure

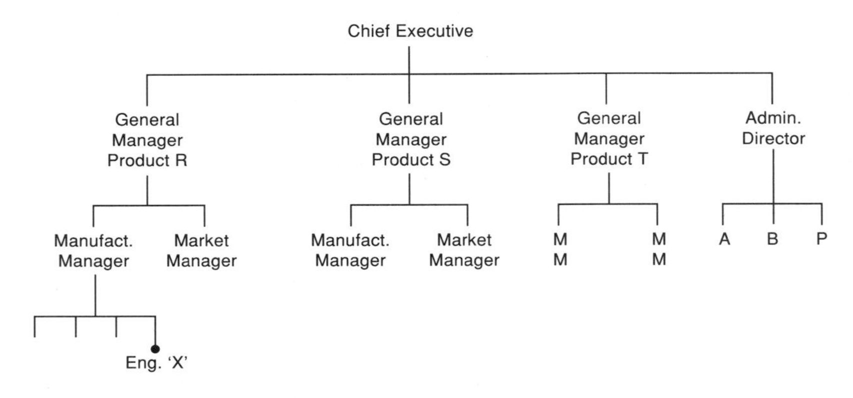

Figure 3.8b Product-based 'B' structure: how effective is the role of Eng. 'X'?

Situation 2

Another problem concerned ownership and communication. Dissatisfaction was expressed by people working at certain levels of the organisation: namely, decision making was remote and reasons for rejecting proposals made were not relayed to individuals or at best were incomplete. Susan's recommendations included options A or B, shown highly simplified in Figures 3.8a and 3.8b. Analyse the effects of these two outline designs on people working at the lowest level and then answer the questions set later.

A typical difficulty brought to light, and unfortunately readily accepted by management, is now given.

Situation 3

A large quality assurance problem needed investigating and solving. The information required will come from different quarters, so a cross-functional group was formed to integrate the findings. This group, when formed, needed a day's training in appropriate problem-solving techniques. The group then spent three days on the problems identified.

These activities need to be specially approved and budgeted, but who will pay? Production did not want to because it is training and involves people from other areas of the company. Training personnel do not like this idea because it is not strictly a course. Decisions should be made on cost-effective grounds but this can ignore some of the bureaucratic realities of departmental budgeting. Situations like this demonstrate how the organisation of responsibility allocation can become unresponsive to the needs of the company. Everyone wants to know When?, so the How? is often ignored.

What would you do in these circumstances to remove the costing problem of the training and other necessities? See the question set later?

E. Questions for Reader and References

For the three situations uncovered by Susan's team, analyse them from the position of the alternatives possible.

Situation 1

1.1. Change the flat organisational structure by reducing the 20 posts and form a combined pyramid of communication to prevent arguments and which identifies ultimate responsibility.
1.2. Many designs are possible but find at least two different options and make notes on your reasons for each combination.
1.3. What particular problems emerged from this exercise?

Situation 2

2.1. The two designs of structure shown are operationally different. Analyse the differences as applied to the engineer, 'Eng. X', as he attempts to do his job.
2.2. Use the potential problem of needing £50,000 to invest in new equipment but being faced with the situation that only £70,000 is available for the whole company in this current year. Consider 'Eng. X' trying to get a favourable decision in structure A with the Manufacturing Manager holding the £70,000.
In structure 'B' he wants to impress the Chief Executive who is the one who holds the £70,000 budget. Analyse what may happen to 'Eng. X' as he tries to progress his ideas through each organisation system.

Which structure provides the best opportunity for him to be successful? Give reasons for the chosen structure.

2.3. Analyse also the role of the Chief Executive in each of the two structures. How will his actions and style be affected by each design?

Situation 3

3.1. Outline a system to get round the departmental infighting over budgetary control.

Overall Question to Consider

'Most of the organisations that I am familiar with are constantly reorganising'. Do you agree or disagree with this statement? Give reasons for your reply.

Suggested references for further information

Production and Operations Management, D.M. Fogarty, T.R. Hoffmann and P.W. Stonebraker, South Western, USA (1989).

Concepts of Strategic Management – Formulation and Implementation, L.L. Byars, 3rd edn, Addison-Wesley Longman (1991).

Strategic Management for Decision Making, M.J. Stahl and D.W. Grigsby, PWS International Thomson (1992).

Production and Operations Management, A.P. Muhlemann, J.S. Oakland and K. Lockyer, 6th edn, Pitman Publishing (1992).

F. Proposed Options Put Forward by the Consultative Committee

Situation 1

Reduction of the multi-involvement of the Managing Director with each of the 20 sections of the organisation could be attempted by two fundamental designs: (i) product based or (ii) function based.

In Figure 3.9 each of the designs is outlined; they are not written in tablets of stone and can be readily modified to suit the Managing Director's wishes.

(a) Product based is clear cut and identifies the main reason for existence by channelling people's attention. However, considerable duplication of functions is necessary but this should not prevent co-operation and communication between the relevant specialists.

(b) Function based concentrates on the role of the specialists. From this specialisation, attention to the five products is reinforced from a position of strength. However, if the specialisation involves many experts it is not unknown for informal compartments to be created for each product and the advantage can be lost.

The particular problems identified by Susan's team included staff demotion and the need to combine similar roles; also redundancy or 'golden handshakes' may be essential.

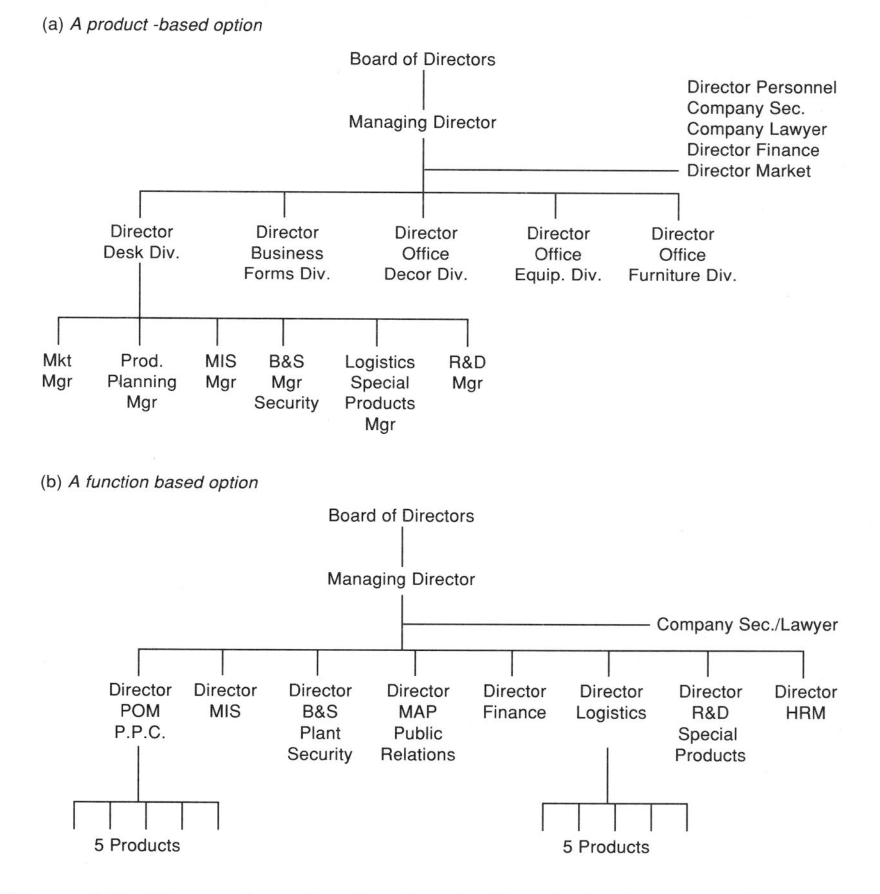

Figure 3.9 Two options for situation 1. Changing the structural design by adding an extra layer of management makes a fundamental difference through decentralisation.

Situation 2

Function design

The available £70,000 in the hands of the Manufacturing Manager means 'Eng. X' is only two steps from the final decision. If he can convince Production

Manager R of the need for £50,000 then Manager R and hopefully 'Eng. X' can attend with S&T management the meeting that makes the final decision.

Product design

This time the £70,000 is controlled by the Chief Executive, who is three steps from the investment recommended by 'Eng. X'. Extra levels will guarantee that he will not be at the meeting that makes the decision. His boss may be but the detailed knowledge in depth he possesses is lost to the decision process. The three product General Managers may agree a compromise on the £70,000 which will mean that £50,000 for product R will not be allocated. Hence 'Eng. X' may prefer a Functional Design organisational structure.

From the Chief Executive's point of view, the Functional Organisation means that he can delegate into specific areas and deal with any identified problems, generally through the appropriate director. In the case of Product Organisation he may have to arbitrate between General Managers who cannot agree about, say, a marketing problem solution because of the vested interests involved. Factors like this could leave the Chief Executive feeling that he has to make all the decisions. A level of management involvement appears to be missing if personalities clash and the interests of the firm are secondary.

Situation 3

The necessary money can be made available in many ways but some popular answers include:

(a) Give a budget for change and development to selected managers under the heading of continuous improvement. In this case the Quality Manager could finance a number of investigations and then justify them by cost savings or customer satisfaction.
(b) The Managing Director could have his own continuous change budget and he decides what will be financed when departmental managers make a bid.
(c) All investigations could be listed at a joint meeting when the contribution from each manager's budget would be agreed. This money is then immediately put into a project budget under the control of the person responsible for the investigation.

Summary

The combination of organisational design criteria with employee attitudes and modes of operation is difficult to predict. It is not an exact science by any means and what works in one place may not work elsewhere. This is why the desires of the Managing Director are paramount, because he is re-

4 How Long to Make and Deliver?

Section

A. Learning Outcomes

Is management really in control of a situation? This is a fundamental question that can be analysed in may different ways, but essentially the criterion is to establish whether what the management wants is practical and then finally whether this is achieved. A simple measure of success is schedule adherence, namely, is the required planned schedule produced to a success rate above 95 per cent?

This case study will demonstrate that:

1. Releasing work into a manufacturing environment and putting completion dates on the documentation do not guarantee success or create control conditions.
2. When completion rates are poor, the act of releasing more work into the area merely exacerbates the problems.
3. The more work in progress created as a result of poor schedule adherence hides many problems, and management then relies on 'fire fighting' as a partial solution.

4. Lack of knowledge concerning activities on the shopfloor is the clearest pointer to production not being under control.
5. Some output can be achieved in any system, but the costs of work in progress and the unpredictability of which jobs constitute the output in any period can be a lethal combination against the survival of the firm.

B. Learning Objectives

After studying and analysing this case study the reader should be able to:

1. Explain the importance of decision criteria to enable the successful throughput of work.
2. Identify key features of a production control system.
3. Define capacity and scheduling needs to allow for a high degree of success.
4. Understand the problems of working in an environment in which the failure to achieve a plan is the only feedback to the workforce.
5. Differentiate between the short-term and long-term needs of production control.

C. The Basics of Production and Control

This activity varies according to the type of industry. A main element that helps set the scene of problems in this area is the P:D ratio. This is defined as

$$\frac{\text{Production lead time } (P)}{\text{Demand time from customer } (D)}$$

If this ratio is less than one, the production control system is working under advantageous conditions because the available time to produce from the order is less than the time that the customer is willing to wait. This gives flexibility for management to make strategic decisions concerning all their customers. See Figure 4.1.

If this ratio is greater than one, the problem arises of how to supply the customer on time when time is inadequate. Generally, the techniques of forecasting, stock build-up, modular design or subcontracting can be used to reduce the full effects. Figure 4.2 shows the problem graphically.

This is a difficult type of industry to work in if customers and shareholders are to be satisfied, because of the costs associated with the inevitable errors of the techniques used.

The majority of companies work with a P:D ratio greater than one, especially when supplying for customers that use Just In Time techniques, and competition is then centred on delivery achievement. Companies can be

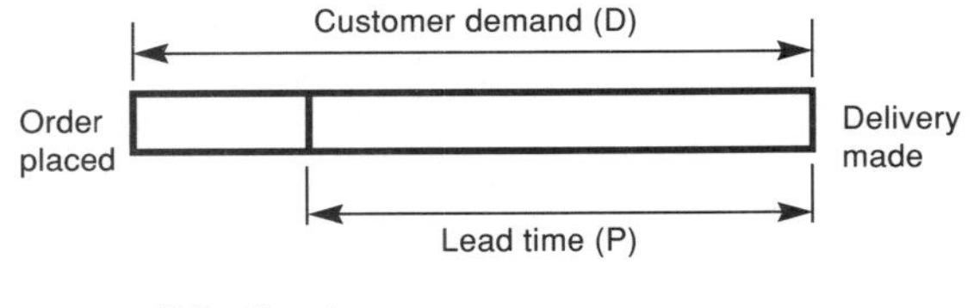

Figure 4.1 Control is easier

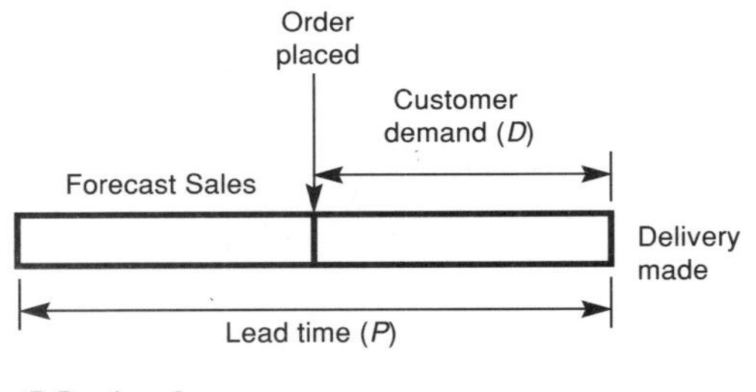

Figure 4.2 Control is demanding

classified by their *P*:*D* ratio on various important components/assemblies they produce; for instance, a company with a value of 0.5 should have little trouble in satisfying all customers needs, while a company with a *P*:*D* ratio of 4 will need to be brilliant at forecasting to be successful. In the long term, every effort must be made to produce a manufacturing strategy to get this value down to something approaching one.

Another factor involved with the ease or difficulty associated with production control can be outlined by three familiar terms into which components/products manufactured are sometimes classified:

(i) 'runners' (product made all year),
(ii) 'repeaters' (products made on a regular intermittent basis), and
(iii) 'strangers' (products never or rarely seen before).

A company that has a high level of runners and repeaters has a more predictable output and knows how to optimise the situation. Just In Time and Kanban techniques thrive in these companies.

High levels of repeaters with a few strangers can also mean that many of the advantages mentioned above can be obtained.

However, high levels of strangers with few repeaters will highlight the day-by-day difficulties of bringing resources and requirements together in a successful plan.

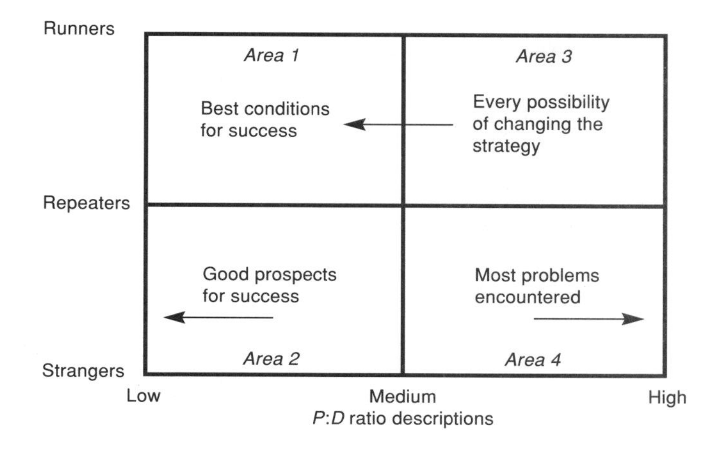

Figure 4.3 Matrix of degree of difficulty inherent in a business

Figure 4.3 shows the effects of the interaction between the two major factors discussed: variety and delivery. Hence the level of Production Control is only a relational quantity, important in all industries, but success is defined differently from Area 4, 75 per cent, to Area 1, 100 per cent, for delivery achievement to promise given, that is for schedule adherence.

Runners and repeaters give a company every chance to maintain an accurate database of processing time, materials, manpower, stocks and distribution. Working from complete and accurate data enables decision making, based on enlightened experience, to be made. Furthermore, the production workforce are trained and familiar with this range of repetitive products.

The problem of production control in this case does not become consistently easy because there are still many uncontrollable factors to prevent success. However, the experiences plus the reporting system result patterns are familiar and give clear advanced warnings in a way that cannot occur when dealing with products classified as strangers.

An important major factor in production control is whether the system used has been designed to 'push' the work through the workshops by scheduling a list of activities, then chasing particular orders to make them move and treating every process as another hurdle to be successfully negotiated without considering the whole production output needs.

Conversely, the alternative is to 'pull' the work through the system by setting up a series of decision rules based on capacity needs and customer requirements. Runners and repeaters thrive in this culture.

Figure 4.4 illustrates the efforts made at the start of the process to push through the customers' specific order, that is all orders are treated independently. This generates queues at each operation leading to Work in Progress,

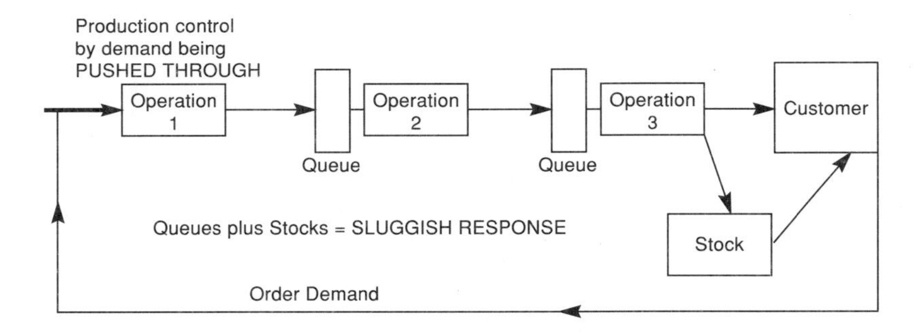

Figure 4.4 The push system

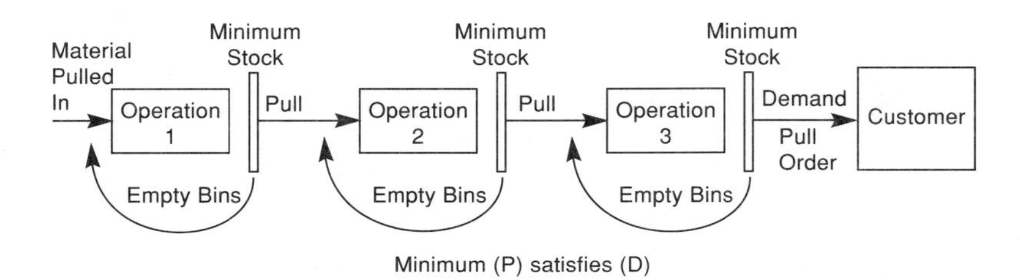

Figure 4.5 The pull system

resulting in delay with cash tied up in partly finished work. If larger batch sizes are used to help provide stock at the stores, this will further slow the system down.

Conversely, Figure 4.5 shows customers pulling off orders from completed or part completed buffer stocks so they are satisfied relatively quickly and the satisfied demand causes a series of 'pulls' through the operations until the system stabilises. The pull system needs some over-capacity to enable it to satisfy the customers' needs, which often vary from time to time.

However, many push systems also have over-capacity which is hidden by the queues and the production of components that are not wanted at present. Under-usage of plant will often lead to cost problems in the eyes of some management personnel, so the conventional push system has the ability actually to produce more than required built into it. This inefficiency, called *utilisation*, is provided for in the costing system and eventually customers may pay for the service they get. Machines costed at x pounds per hour are considered wasteful if not producing something. This makes efficiency and waste part of the system and difficult to identify, because, say, 75 per cent utilisation becomes the standard performance demanded. The alternative 'pull' system only produces components which are required by customers, therefore over-capacity is revealed by inactivity or waiting time. Adjustments can then be made to monitor and minimise this identified waste. One system hides waste and the other system exposes it. Essentially this simple rule applies: if machines are to be kept busy then an excess of manpower is required. If manpower is to be kept busy as a policy then excess machines are required. If customers are to be satisfied then excess capacity is needed. Capacity is very difficult to balance with demand over a period. Capacity can be managed either to increase or to reduce by a series of management actions which are part of production control. These actions include:

(i) Overtime or extra shifts,
(ii) Subcontracting work or whole orders,
(iii) Employing temporary labour,
(iv) Using method study to increase efficiency,
(v) Weekend working,
(vi) Building up stocks in slack order periods.

The previous discussion has been about the structural problems of industry and the basic ways of tackling them. Although computerised systems have not been mentioned, they are not forgotten. Many systems exist but they should be applied to a streamlined manual system and not put in to 'control' the present chaos. Simple solutions tend to work better than complicated ones. Changes need to be communicated to the workforce and production control simplicity helps in this action. Whenever possible, visual production control is preferred; visual control is obtained when the success or otherwise of a productive system can actually be seen working rather than having it proven by a series of computerised printed hard copies of results.

Problems of High Work in Progress

The problems caused by work in progress around a workshop can best be explained using the simply analogy of a bath. The taps are the orders, the drain is the output and the contents the work in progress.

Figure 4.6 shows pictorially what is required in terms of flow and, most important of all, the operating situation. While the WIP (Work in Progress) dominates, there is little connection between the management operating the taps and the drain going to the customers. Figure 4.7 shows that with minimal WIP the management have greater control and the time from input to output is regular and guaranteed. Importantly, lead time is reduced.

The bath analogy demonstrates what is commonly called a 'bottleneck' in production. 'Bottlenecks' (restrictions in production capacity) will always exist because no one has yet established infinite capacity! If potential output exceeds demand then no 'bottlenecks' should exist, but as most customer

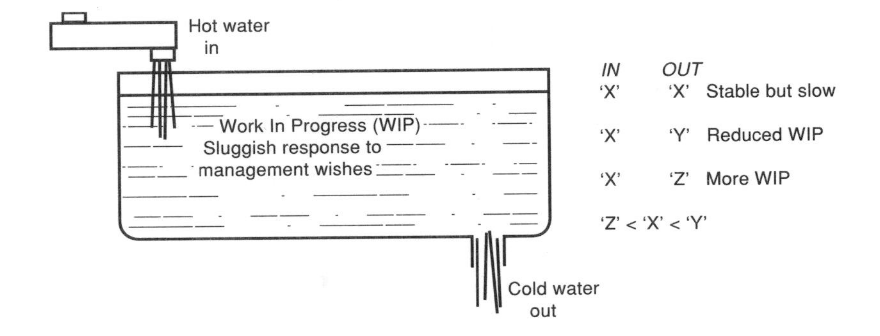

Figure 4.6 Bath tub analogy to effects of high levels of Work in Progress

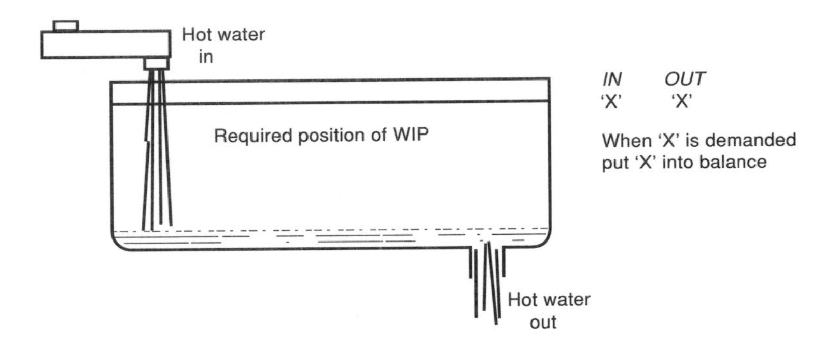

Figure 4.7 Bath tub analogy to effects of low levels of Work in Progress

demands are variable, associated with a variable mix of product specifications, it is inevitable that certain processes will receive a biased over-load from time to time. These 'bottlenecks' are more difficult to deal with because they are temporary. A permanent 'bottleneck' is generally obvious and controllable by overtime or shifts, or even duplication, although this requires more investment.

The simple example of Figure 4.8 explains the fundamental problem of 'bottleneck' occurrence. Say a job needs 5 operations to complete. Each operation runs at different speeds of completion and operation 4 sets the pace at 8 per hour. If the other processes run to full capacity queues will form between 2 and 3, and 3 and 4. If all processes run at 8 per hour then inefficiency and delay are built into this strategy. Can the spare capacity be used? The solution sometimes can be to load a selection of jobs simultaneously on each of the 5 operations and expect that statistically the load will even out. This ignores the slow operational bias of some operations, such as

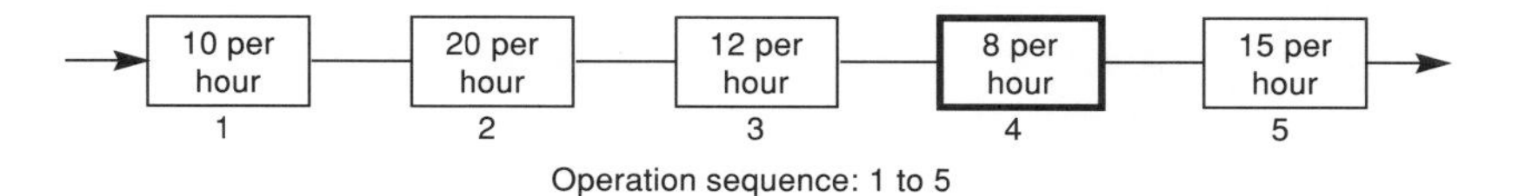

Figure 4.8 Effects of 'bottleneck' in production line

heat treatment, and the fast operational bias of others, such as presswork. Queues are also created at each process, of unknown size.

The real solution comes from 'bottleneck management', that is controlling the throughput at this operation and flexibility managing previous and subsequent operations. This strategy focuses the management attention where it is needed – where efforts in tool developments, overtime and set-up time reduction will all bear fruit. Similar efforts in other operations that already have slack time will just increase their capacity on operations where the demand is not important.

A measure of performance can be used to identify the extent of the sluggishness (long lead time) created by extensive work in progress. The measure of performance is based on a comparison between the estimated time for all the processes needed to complete a batch of components versus the actual time elapsed for this work to be carried out.

Example
10 processes must be carried out on one component and the sum of the time required is, say, 2.5 hours.
10 components are required, hence 10 × 2.5 = 25 hours total processing time. Setting up the batch 10 times at each individual processes takes, say, 5 hours total, hence the grand total is now 30 hours of process time.

In an ideal world with no queues, 30 hours after starting the batch it would be ready for despatch. However, it is more likely, given queues do exist, that the batch takes 10 weeks to be processed because of the delays in the system, so the comparison ratio becomes:

$$\text{Throughput efficiency} = \frac{\text{Total processing time required}}{\text{Total processing time elapsed}}$$

$$= \frac{30 \text{ hours}}{10 \times 40 \text{ hours}} = 0.075$$

or <u>7.5 per cent</u>: this type of result is not uncommon

In practical cases many jobs achieve this low measure ($7\frac{1}{2}$ per cent), while others that are treated urgently and 'chased' through the processes can achieve about 50 per cent.

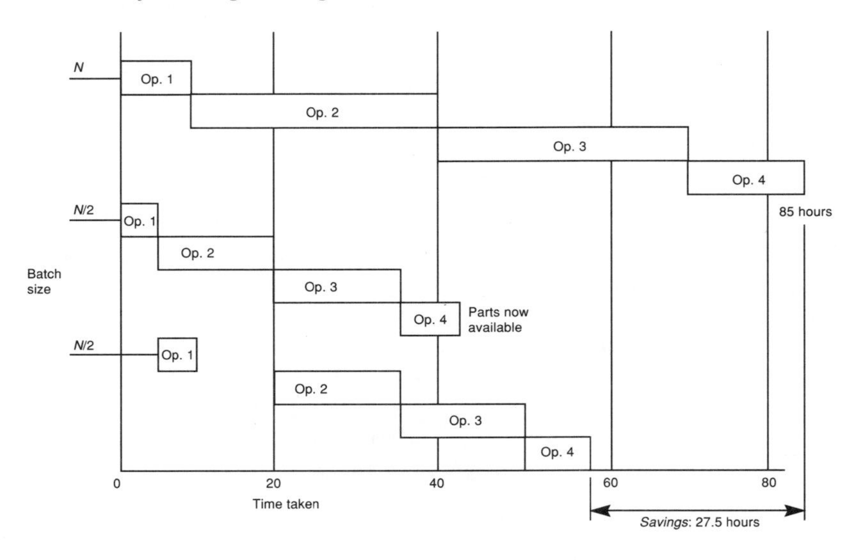

Figure 4.9 Effects of two batches rather than one

These measures of performance attainment sum up the management's production control dilemma; namely with little scope for flexibility because the *P*:*D* ratio is greater than one, how can these levels of performance be accommodated into a feasible production plan? The first response must be to reduce the work in progress and make the system more responsive. Pushing extra work into this situation can only make it worse. It has been known for production controllers requiring, say, 50 components, to have to work so long for them to arrive that they have increased their subsequent order to 200 to create a buffer; however, this is self-cancelling because the wait is even longer and the shortfall delay prolonged for the assembly schedule.

Whenever possible, the production management seeking a more responsive system should consider reducing batch sizes to increase the ability to provide what is needed. Do companies really want to receive 1000 components simultaneously or are they willing to accept 5 × 200 each week for 5 weeks? The second scenario is more likely to be successful than the first because the order ceases to be the occasional large 'stranger' and then becomes a 'repeater' when this policy is planned and pursued.

The effect of processing large batches and the difference made when the batch size is reduced can be shown via the Gantt chart of Figure 4.9.

By reducing the batch by a half and processing two smaller batches in sequence the in-built inertia of large batches is clearly demonstrated. It is this advantage, albeit with more paperwork and management effort, that comes from batch size reduction.

A further problem that comes from WIP build-up is demonstrated by operators having a choice concerning which particular job in a queue is processed

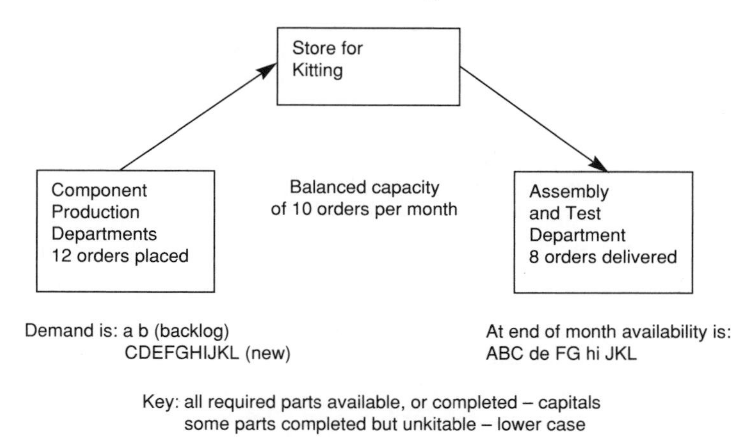

Figure 4.10 Uncontrolled scheduling reduces capacity

next. If any management logic is missing from this choice and if it is based on bonus earned, tool availability, preference, ease of working or mistakes, then even more complications arise. A following example is one way to demonstrate this situation.

A common problem in multiple-part assemblies is requiring all the parts to be available together for assembly of a saleable product to customers. Controlling the production of all the required parts is not easy because so many people make decisions at each stage of output and have little idea of the company needs. By management scheduling more components than available capacity allows, the chances of overall success are greatly reduced.

If each of the assemblies comprises ten items, then if one or more items are not available all the remaining items are kept waiting in the stores (work in progress).

Figure 4.10 shows how the effects of over-scheduling are often made worse by a backlog from a previous period.

To demonstrate this effect, each of the component assemblies ordered is given either as upper case (capital) or lower case (small). Capitals means that all the parts are available for assembly. Small letters mean that one or more parts are missing.

The worst scenario would be working to complete as many parts as possible but not having one complete set available. In such a case, the stores would be full but no saleable products could be assembled and tested. Technically, this would be a 100 per cent reduction in saleable capacity.

A step-by-step outline is now given.

Step 1

The component production department starts processing the operations on 12 sets of parts with the date priority of the individual operations paramount;

this does **NOT** produce 10 complete sets of parts but 12 partially completed sets, as processing is carried out across operations of available work.

Step 2

In the stores the parts that are available are made up into kits for assembly. A shortage list is then made to enable the incomplete kits to be completed so they can be issued to the Assembly and Test Departments.

Step 3

This list is now returned to the Component Production Departments and urgent requests made to provide the parts to complete the assembly kit. Under these circumstances the further disruption of the schedule means that the company may only get 8 orders completed in the next period. Hence a backlog of customers' orders is created.

This scenario will be exacerbated period by period and further disruption will cause even greater delay so that even with the most valiant efforts of the progress chasers, the real effective capacity of 10 will be reduced on a monthly basis until some kind of stability is reached at a very inefficient level. The importance is now established of loading only orders that can be achieved on to a system, hopefully without a backlog of work. Anything that can go wrong will probably do so, therefore the fewer potential chances available for management to lose control the better.

D. Introduction to problems of Jones Tools Ltd

Jones Tools Ltd is a small company employing 30 people making specialised punches and dies for the press tool industry. These components are used to pierce holes and form blanks in sheet metal when positioned in tool sets.

There are a number of problems, mainly relating to customers' expectations concerning delivery response to urgent orders. The specialised nature of the materials, shapes and sizes of punches and dies produced, means repeat orders are few and well spaced. This is a customer-driven business. In the past the company would deliver completed tools to customers 8 to 10 weeks after the order was placed. Market conditions now dictate that customers want a 4 to 5 week response, and some competitors are easily achieving these targets.

Jones has a reputation for quality which is keeping most of their customers loyal, but know that it is losing some orders when the work is really urgent. This trend is well established, which means the long-term prospects are not good unless the delivery achievement is dramatically improved. Different plans and systems have been tried, without success, to speed up the flow of work through the workshops. At one time a computerised management production package was purchased with recommended software. This was bought in haste and could not be fully used because the database of the company was too inaccurate and incomplete. These data problems are the result of mainly new and specialised orders which have variations from previous work necessitating different methods and resources to be applied. The timescale for these changing resource demands was estimated, which produced only limited success, hence throwing out-of-phase the interactive plans and schedules.

The inaccuracies of the database led the company management into using a basic financial planning tool to decide delivery dates. From past experience the value of throughput obtained in a week was established. This was taken as a measure of capacity, hence the value of each order obtained was compared with the previous financial commitment and promises for delivery given when and where the order value could be slotted into the budgeted capacity. The financial target, if achieved, resulted in the planned profit for the company. This system did work from one point of view but failed to improve the situation; it only perpetuated it, that is it increased work in progress. The target in pounds had been set against an inefficient system, so even when this target was occasionally exceeded it was difficult to assign reasons for this occurrence which may have helped decisions in the future. Obviously, a more detailed feedback of results was needed.

The management of Jones was aware of a number of symptoms that caused delay to work in progress. These included:

1. Batch sizes of 8 to 16 for punches and dies were above the order quantity, which always included spares in anticipation of breakages and wear.
2. Queues at the first machines in the operational sequence arose from overloading or under-achieving targets in previous weeks.
3. Each queue formed in the workshop at a machine meant an in-built delay to prevent the movement of work, to achieve a four weeks' target for delivery. See the 'bath' example of Figure 4.6.
4. The specialised nature of the work meant that complex shapes needed about 10 separate operations to complete the component. This gave opportunities for 10 queues to form for each batch of components processed. These opportunities had been taken, unfortunately.
5. Morale in the workshops was very low because of the apparent continuous failure to produce to the set targets.
6. This produced a culture of failure, and new ideas and changes were not well received, especially if they were complicated, short-lived and did not improve the situation to any degree.
7. Each batch of 'urgent' components specially treated by management to jump over many of the process queues created further delays in product queuing.
8. Cash flow problems resulting from inefficiency caused the opportunity for large-scale investment to disappear.

9. The company had a *P:D* ratio of 2, with no chance of producing stock to forecast because of the variable nature of the order specification sizes and shapes.

It is from this position that the Managing Director requested the Group Eng. Services to investigate and make recommendations over two days. (This is the most consulting time they can afford. Service time from Group HQ had to be paid for from each autonomous unit.)

The Present Position in Detail

The consultant, Frederick Downes, came for the two-day period and observed many of the difficulties previously outlined. He considered the most important facts were:

(a) The congested nature of the workshops because of queues of work in progress; these cause most of the problems and must be dramatically reduced.
(b) A plan to produce orders inside the 4 weeks' delivery requested must be found or the company would drift towards failure.
(c) An analysis of the planned operation processes needed for a large sample of work showed that the average number of production steps was ten and that the initial two or three processes took the longest individual times.
(d) Queues existed at each process so it was difficult to identify the causes of slow throughput. Normally a very large queue builds up at one particular process. This expected 'bottleneck' would have to be identified by other means.
(e) The present method of regularly loading each week with a new set of orders, while a large backlog of work on the processes still remained, must stop.
(f) The workshop must become responsive to management control and direction on all parts, not just a few.

Jones Tools Ltd required a variety of processes to produce the tools; these included turning, milling, drilling, internal and external grinding, honing, various heat treatment processes, electric discharge machining, presses and saws. The workforce tended to specialise on one or two particular processes. This had not caused trouble in the past because of the work in progress buffer, however a problem could be seen in that the rigidity of this system would prevent flexibility when conditions changed in any new plan. A flexible workforce would be expected to tackle a larger number of the processes that the company used. This would involve time and training.

E. Questions for Reader and References

Consider the Possibilities

1. How many different possible methods could be listed for the management to gain control of the situation?
2. Can new decision rules be devised that will establish priorities and reveal the essence of the problem?
3. Explain the basis for the recommendations reached after analysis.
4. Outline a stepped plan to enable the management to satisfy the customer requirements.

Commit the ideas considered to paper so they are reasonably structured and can be tested. Experience shows that random thoughts, with no permanent record of analysis, tend to be self-serving and uncritical of performance.

Suggested references for further information

Manufacturing Systems, edited by Victor Bignell *et al.*, Blackwell (1985).
Production and Operations Management, D.M. Fogarty, T.R. Hoffmann and P.W. Stonebraker, South Western, USA (1989).
Operations Management – Concepts, Methods and Strategies, M.A. Vonderembse and G.P. White, 2nd edn, West, USA (1990).
The Visual Factory – Building Participation through Shared Information, M. Greif, Productivity Press, USA (1991).

F. Proposals Adopted by Jones Tools Ltd

Frederick Downes realised that a new system using the existing documentation was preferable, so the first planned changes had to be simple to reveal the underlying problems. After such a short time (2 days), which gave no opportunities to monitor the effect of changes over weeks/months, except by occasional telephone calls, he knew he had to recommend a 'shortcut' procedure to the firm.

The basis of his recommendation originated from the performance that the company wanted, that is 10 processes completed in 4 weeks (20 days' activity). It would therefore be necessary for all work to move from one process to the next in 2 days. This was to be the target: providing a decision rule that chose which orders were selected for processing, that is scheduling each operation locally. Those that had been two days in the process queue had to be identified by the operator, and management informed about the delay.

This change would be of little use for the present system because identifying where many of the late orders were was not an end in itself. A decision had to be made to stop loading the workshop with orders until no identified late work (over a two-day target) existed at the three early operations.

Queues were to be separated at each operation into those under the two-day deadline and those that had exceeded it.

A visual appreciation of the forward load on each operation would then be available on a continuous basis. This hopefully would be the clear result. The movement marker for each operator at each operation was to be the route card. This document had details of each process necessary to fulfil the order, but the time estimation for any one process was sometimes suspect. By using space available for date, time and operator signature the pace of movement would be ascertained. As the queues reduced it was expected that the two-day exceptions would follow suit. Any left in one place in particular would demonstrate the 'bottleneck' potential of that process. As this process would be isolated and differentiated from the other processes, management could direct its efforts exclusively at these exceptions, while the bulk of the orders moved on time through the rest of the workshop. Those operators that were skilled in more than one operation could be moved to the overloaded processes to give immediate help. The company did have an excess of process plant to workers, however all processes were not duplicated.

Finally, Frederick Downes, in his report noted some long-range strategic aims for the company as the needs became known:

1. All workers to operate at least three types of plant to enable flexible operation of processes according to demand.
2. Data to be collected on a range of tool shapes and sizes so that standardised synthetic data could be built up to make the time estimation more accurate and eventually capable of being put into an operational database. This base could then be used for simulated estimation times to produce basic shapes of orders required.
3. A control system of colour codes was needed to identify delayed orders so that the visual monitoring of the progress of each job would be enhanced.

Results of Proposals Adopted

The short, quickly produced report was accepted by half of the management (2 out of 4); the reservations made by some were negatively opinionated and could only be disproved by experience.

One week later the plan started. The first step taken was not loading any more orders into the workshop, and then getting workers to note when the bulk of their work moved inside the set target of 2 days. Stabilisation only took 4 weeks and resulted in an 80 per cent achievement of customers' new requirements. The main reasons for the 20 per cent failing were inaccurate estimation of times to complete a component and a bottleneck of parts at one particular operation. The previous performance was around 10 per cent achievement.

The 'bottleneck' operation concerned was the electric discharge machining of die moulds. This is often one of the last operations to be performed. The irony of this 'bottleneck' was that the company had three such machines until recently, when one was sold to another company because of lack of use. This poor utilisation came from the delays prior to this operation, causing only a partial flow of work to the three machines. Now the delays were minimised the true 'bottleneck' was exposed. Two shifts on each machine, including weekends, corrected this major delay over a three-month period.

The net effect on the company was better service to customers, more orders and a better cash flow position. Eventually the Managing Director wanted to 'buy in' more expertise from HQ to effect further savings now the survival of the company was assured.

Summary

When a company is in difficulty to a degree that 'fine-tuning' of a problem does not produce the necessary results, then drastic evaluation is needed from a new perspective.

This case study shows that it is sensible to start with the required outcome and then determine the system and resources necessary to accomplish it. An overview is required that ignores entrenched beliefs and protection of position. Once the message is communicated of how exactly success is to be achieved and the continuous commitment to see the plan through is demonstrated, albeit only by two out of four managers, then a culture change is possible.

5 Costing to Aid Decisions

Section

Costing to Aid Decision Making at Waynes Scales Ltd

A. Learning Outcomes

The importance of the collection of useful and accurate data for analysis, from which good decision making is made, cannot be over-emphasised. Having the correct reliable inputs to any costing system is the basis upon which insights are obtained.

A costing system can be thought of as a model that replicates the real world when it functions correctly. A good costing system predicts the effect of changes so that cash flow advantages can be made; for instance, a wage award of 5 per cent can be simulated and the full effect on profitability established to within close limits, and similarly with improvements in sales output.

This case study will demonstrate that:

1. Overhead or fixed cost assimilation into product cost is difficult and can often mislead the decision maker.
2. As the drift from labour cost to investment cost continues, different ways of cost modelling need to be tested.
3. As the company develops, the costing system needs to react to the new situation by itself changing.
4. No costing system is written on tablets of stone, so regular testing and challenging of its suitability of purpose are paramount for good decision making.
5. An easy costing solution is not always available but certain dramatic pitfalls that do exist can be eliminated from costing analysis.
6. A sceptical approach to number generation by a costing system is required, to prevent too much trust being placed on existing methods.

B. Learning Objectives

After studying and analysing this case study the reader should be able to:

1. Explain the problems of overhead absorption.
2. Identify critical stages in a build-up of final product cost.
3. Define and understand marginal costing philosophy.
4. Differentiate between fixed and variable costs incurred.
5. Analyse different costing problems in a critical fashion and not just think of arithmetical answers.
6. Understand the connection between information and good decision making which eventually is audited for degree of success.

C. The Basics of Costing Products

Although costing can seem arithmetical in nature and complex in its execution, it is not only a science but also an art. The science part is the systematic analysis of data collection while the art input is from philosophies, strategies and decision criteria based on subjective thoughts involving common sense. These two strands of costing will be identified throughout this chapter. Costing is the activity of *explaining* what will happen or has happened to the money flow as the company operates.

All companies should know their expenditure accurately and also their sales figures, so that the difference between them shows clearly how well the company has performed over the last year, quarter or month.

However, these figures only give the result, not the background reasons for the result. Costing systems are designed to explain why the result occurred as it did. This information should enable the management to make informed decisions concerning the operation of the business and hence have more financial success. The alternative method is based on having no explanation and using a subjective analysis for the likely decision options for management. This effect can be illustrated using a simple example with basic round numbers.

Consider two companies A and B: one with a costing system (B), one without (A).

Results for year 1

	Company A	*Company B*	
	(£)	(£)	
Sales	100,000	100,000	
Expenditure	80,000	80,000	
Costing system	—	10,000	
Profit	**20,000**	**10,000**	+ **Explanation**

The starting point is similar for A and B, but money has been invested for information in the case of company B. The expected outcome is that better decisions are made in the light of the information, which enables a different scenario to be made for B compared with A, since A still has an unenlightened management.

Results for year 2

	Company A	*Company B*	
	(£)	(£)	
Sales	110,000	120,000	
Expenditure	100,000	80,000	
Costing system	—	10,000	
Profit	**10,000**	**30,000**	+ **Explanation**

If the costing system cannot pay for itself in terms of performance then it is better not to pay for this information. A number of smaller companies often do make this decision and have to live with it. If their product line is simple then it is possible to survive, but complicated production systems create real problems without costing information.

The first step in the creation of a costing system is to categorise costs/expenditure on items as either *fixed* or *variable*. Some sources use the terms *indirect* and *direct* costs to establish the system.

Fixed costs (*indirect costs*) are those which remain stable over a year even though the volume of manufactured output is variable.

Variable costs (*direct costs*) are those costs which change in direct proportion to the volume of manufactured output. An exhaustive list is not supplied here, but an indication of which costs are normally allocated to each class is given. Some companies faced with a problematic decision call a few costs *semi-variable* and split them into their fixed and variable elements.

A simple example of these costs, according to some companies, may be the telephone costs. The rental is fixed, but usage varies with activity.

Typical cost allocations

Fixed costs	*Variable costs*
Rent	Material (various forms)
Rates	Power
Heating and lighting	Labour costs
Insurance	Expenses
Canteen costs	(i) Subcontract
Directors' salaries	such as Heat treatment
Administration costs	Plating
Designers' salaries	Galvanising
	(ii) Consumables

The contentious item listed is the Labour costs. If the manpower can be employed by the hour and sacked by the hour then it is truly variable. However, if specialised skills are needed and redundancy and replacement costs are considered, then the direct worker is just as secure as the designers working in the offices, categorised as Fixed costs. Seasonal and temporary workers would be seen as variable because they can be used to accommodate trends in demand. Note the upsurge in these classifications recently. The two lists define the problem facing cost accountants. One of them (variable costs) can easily be linked to products or services and this results in the correct allocation of expenditure to each product. The other class (fixed costs) has no obvious association with the products or services achieved and this results in allocations often being made based on criteria that sound reasonable but can be riddled with anomalies and distortions that cause errors in decision making. A number of different allocation methods are available, and are in regular use; these will be analysed later. However, the problems faced by management can be demonstrated with a humorous little story that has a grain of truth in it.

The Managing Director of a company making only one product assembly considers last year's results. He knows the company spent £100,000 in total and made 100 assemblies. He concludes that it costs £1000 to make each one. His salesman, knowing that the firm has spare capacity, sells another 20 for £900 each. The Managing Director is furious when he finds out and sacks the salesman for selling below cost price. Is he right to do so? Let us consider the case. For ease of calculation, say that the £100,000 expenditure is made up of £50,000 fixed costs and £50,000 variable costs. Hence, 100 assemblies cost £50,000, that is £500 each in variable costs, so the £900 provides £400 extra cash flow for each one sold. The fixed costs do not increase and have been paid off by the previous 100 sold. It is now clear the company in using its spare capacity has made 20 × £400 = £8000 profit, not a loss as originally thought.

Happy ending – the salesman was given his job back with a suitable bonus. The example shows the problem of using average costing which will change when more or fewer items are produced or serviced. The Managing Director's dilemma is repeated in more complicated ways in many organisations. The rationalisation of fixed and variable costs is at the root of many costing decisions.

In Figure 5.1 the activity of overhead distribution as a philosophy is demonstrated. However, it is more demanding than the theory suggests because, where money is concerned, the accuracy of the outcome is paramount to management.

Sales = Expenditure + Profit
Expenditure = Direct costs + Indirect costs (fixed)
Cost of products = Direct cost proportion identified + Allocated overhead (OH) decided by criteria chosen

The distribution of the proportions of the fixed costs are dependent on the criterion chosen. Decision-making effectiveness is greatly affected by a chosen system. A straightforward example is designed to show the effects of allocation on decision making.

The Foundry Quotation Problem

A component to be cast from iron is required in large quantities and a quotation is required from the foundry manager. He can specify one of two processes in the foundry and base his delivery date on the availability of the process capacity. The two processes are either Manual Hand Moulding to prepare moulds for pouring molten metal into, or Machine Moulding ready for pouring. The data available initially are as follows:

Hand Moulding	*Machine Moulding*
Labour content 75p 100% Allocation of overheads on to labour cost	Labour content 25p 600% on allocation of overheads on to labour cost

Material usage the same.

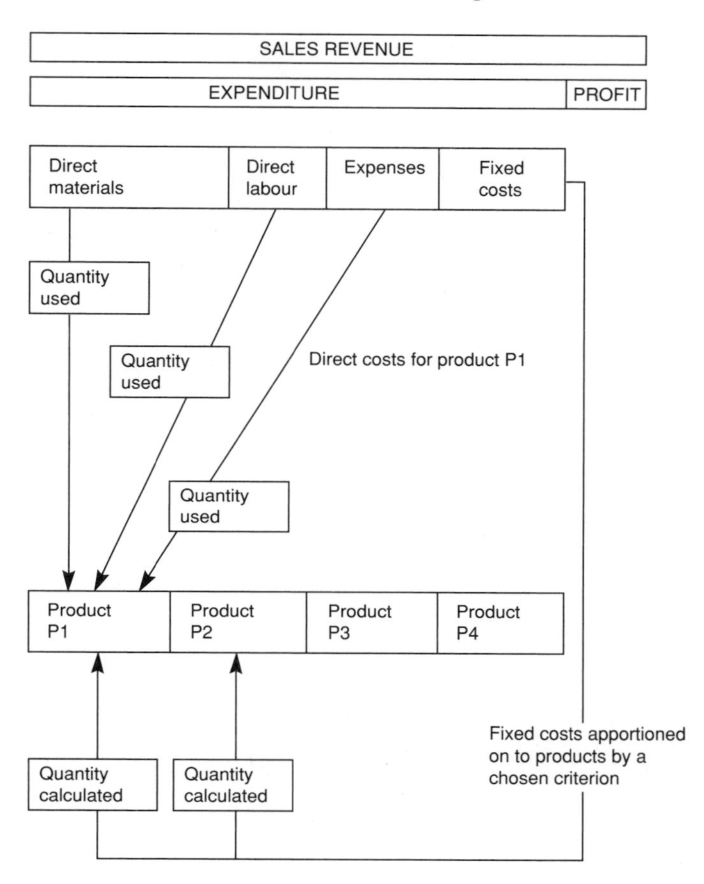

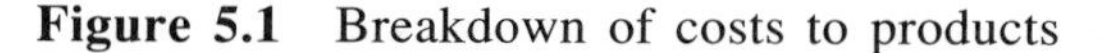

Figure 5.1 Breakdown of costs to products

Calculation

Hand Mould	*Machine Mould*
Costs = Labour + OH	Costs = 25p + 25p(600%)
= 75p + 75p(100%)	= 25p + 150p
= 75p + 75p	
= **150p**	= **175p**

The inevitable conclusion is that they will hand mould this very large order on a cost basis!

The manager is unhappy with this conclusion and obtains further information, which is: Hand Moulders earn £6 per hour, whereas the Machine Moulders earn £5 per hour. The fixed costs that can be clearly attributed to each process are £3 per hour for Hand Moulding and £10 per hour for

Machine Moulding. With this new information the manager establishes the following:

Hand Moulders	*Machine Moulders*
Wages of £6 per hour	Wages of £5 per hour
gives $\frac{£6}{0.75}$ = 8 components per hr produced	gives $\frac{£5}{0.25}$ = 20 component per hr produced

Comparative costs = Labour + OH per component

	Hand Moulders	Machine Moulders
	$= 75p + \frac{£3}{8}$	$= 25p + \frac{£10}{20}$
	$= 75p + 37.5p$	$= 25p + 50p$
	= **112.5p**	= **75p**

Hence, the conclusion is to use Machine Moulding and have 37.5p advantage 20 times per hour, and satisfy customer deliveries faster. Time taken is an important factor in costing. Percentage allocations often ignore this important fact when only individual costs are quoted.

The allocation of fixed costs to products is now examined specifically on the basis of matching criterion to process.

A number of methods of fixed cost distribution are used in different industries; each one is now briefly examined for effect. All are based on the previous year's results (historical costing) and it is assumed that the proportions of cost will be similar year on year.

The Different Criteria

1. Percentage allocation based on labour content

Traditionally, this is a very important and popular method because manpower is seen as the creator of wealth and hence the cost-driver. The total fixed costs for the previous year (overheads) are divided by the wages paid to direct workers to form a ratio expressed as a percentage:

$$\text{\% on direct labour to recover overheads} = \frac{\text{Last year's overheads} \times 100}{\text{Last year's direct wages}}$$

The resultant rounded figure is used to apply the fixed costs to all work undertaken during the following year.

Example

3 hours of direct labour at £6 per hour and an overhead cost of 400 per cent results in:

Production cost = Direct cost + Overhead apportioned
= Materials cost + Labour cost + 400% of labour cost
= Materials cost + (3 × £6) + 400% (3 × £6)
= Materials cost + £18 + £72
= Materials cost + £90

This method may have some legitimacy for activities such as assembly and tool making which are labour-dominant, but using the system for direct workers involved in process control and machine operation is questionable (see foundry problem). It also suffers from the problem of differential wage rates. If someone on £5 per hour were to do the same job it appears that the total cost becomes Materials cost + £75. The larger the percentage added to direct labour, the larger the error. Further, if the percentage is a blanket one for all activities, is this fair on the lower-cost and easier activities of the company?

2. Percentage allocation based on material cost content

This is an occasionally used method for industries often dominated by material content, such as Foundries and Press Working Departments. The structure is similar to the previous method and gives

$$\% \text{ on direct material} = \frac{\text{Last year's overheads} \times 100}{\text{Last year's material usage}}$$

Example
Material content £2 with percentage recovery of OH on material: cost = 600 per cent.

Manufacture cost = Material + Overhead (which can include labour charges which are relatively small)
= £2 + £2 × 600%
= £2 + £12 = £14

If a cheaper material is used because of a specification change, say £1.50, then the cost above becomes:

Manufacture cost = £1.50 + £1.5 × 600%
= £10.50 Has £3.50 been saved or £0.50?

Does this help us to make correct decisions?

Situations like these can create a culture of material cost consciousness which is stronger than it should be. The end result is that next year the ratio will have changed because the overall divider (that is, materials cost), is lower. Recovery rate rises to 650 per cent and the emphasis to cut materials cost is stronger.

3. Overhead recovery on units made

If a company has a limited range of products that are similar to one another, such as Refrigerators, Vacuum Cleaners or Tonnes of Steel, then they can be considered to be either Units of Production or Weighted Units of Production. Weighted-units ideas are sometimes used for larger units (non-standard) and they are considered to be worth, say, 1.2 Standard Units rather than one. The total is then formed by the weighted average.

$$\text{Overhead recovery on each unit} = \frac{\text{Last year's overheads}}{\text{Last year's output (units)}}$$

Example
Overhead recovery = £25/unit; product 'Z' unit weighting 1.5. Then overhead allocated becomes:

£25 × 1.5 = <u>£37.50 for each unit of 'Z'</u>

As this is only applied to straightforward production ranges, the feasibility is stronger. But some units are easier to sell than others and some make more profit, so a change in the weighted mix could lead to under- or over-recovery of overheads (see later notes in this chapter).

4. Overhead recovery by Machine Hour Rate (MHR)

This is a popular method when machines, processes and equipment are the dominant factor. The method is based on a Cost Centre, in this instance a large machine tool, to which the expenses of the company are identified for providing a service to enable it to function. Costs are allocated on a yearly basis to the cost centre when identified as necessary for the centre's activities. Hence costs of Space, Heating and Lighting, Insurance, Maintenance, Business Rates, Power, Labour, Supervision, Production Control, Depreciation of Equipment, Quality and Administration are all added to form a total for the Cost Centre. This figure is divided by the expected available hours of operation in one year to give a cost per hour of operation:

$$\text{MHR} = \frac{\text{Total costs identified for equipment}}{\text{Total hours equipment available}}$$

Example
If MHR is £20 per hour because of £40,000 costs and 2000 hours availability then the system is in balance. If, however, the equipment is only used for 1800 hours then the 200 hours of shortfall means that 200 × £20 = £4000 is under-recovered and is taken off the company profit margin. Con-

versely, if equipment is used for 2200 hours through overtime then 200 × £20 = £4000 is over-recovered and is added directly to the company's target margin. This is the basic weakness of all the allocation methods. Is one year's activity really the same as another?

A further complication arises with the way depreciation is determined at a particular company. There now follows a factual account of the way two companies located 6 miles apart dealt with exactly the same machine tool making similar components:

Company 1 decided to recover the cost of the £60,000 machine tool over 3 years, working the machine 1600 hours per year. Hence the capital cost part of MHR worked out at:

$$\frac{£60,000}{1600 \times 3} = \frac{£60,000}{4800} = \underline{£12.50 \text{ per hour}}$$

Company 2 decided to recover the costs of the £60,000 machine tool over 5 years, working the machine tool 2200 hours per year. Hence the capital cost part of MHR worked out at:

$$\frac{£60,000}{2200 \times 5} = \frac{£60,000}{11,000} = \underline{£5.45 \text{ per hour}}$$

The £7.05 difference can be crucial in competing for similar work. After all, competitors are not likely to be any quicker and regain the difference by efficiency. This subjective input to costing is again a problem with allocation methods.

From the previous four methods outlined, it can be seen that allocation can seem logical and attractive but some have doubts concerning authenticity. Another way of avoiding the pitfalls of allocation is needed to enable decision making to be undertaken with more confidence.

Marginal Costing

This methodology gives management a chance to avoid allocation problems. In calculations, the classic profit formula is often expressed as:

Selling price − Cost to make = Profit

It has already been shown that the 'cost to make' includes an overhead allocation and is subject to the problem of under- or over-recovery of overhead, that is fixed costs, which depend mainly upon the unlikely repeatability of last year's performance.

In contrast, marginal costing is defined as the extra costs incurred to make one more component, or the money saved when one component fewer is made during a period of time. This creates a formula that does not contain overhead allocation as a component, and hence produces a contribution, not a profit:

Selling price − Marginal cost = Contribution

The idea of a *contribution* is that the excess money generated is a payment towards the overheads incurred. When the company's overheads are completely recovered, the contribution then becomes the profit and is obtainable thereafter so long as the marginal costs do not change. Hence the marginal cost is the additional cost of making one more unit, and this will include Materials, Power, Expenses and Labour content (bonuses). To prevent sophisticated detailed calculations being required, a simple alternative is adopted which is useful and easy to establish. Marginal cost is very similar to variable cost. Hence the formula becomes:

Selling price − Variable costs = Contribution to overheads or profit

The advantage of this approach is that each product is considered as a contributor towards the company goal, some providing more money than others.

Example of comparison: allocated versus marginal costs
Three vacuum cleaner models are compared using percentage on direct labour overhead allocation method with a value of 450 per cent:

Individual cost	*Cylinder*	*Special upright*	*Dome*
Direct wages	9	18	6
Direct materials	18	15	24
Direct expenses	9	5	5
Prime cost	36	38	35
Factory OH	40	80	27
Sales dist. & admin.	6	6	6
Total cost	82	124	68
Selling price to wholesaler	90	100	80
Profit	8	(24)	12

Does this mean that if one of each model is sold, the company loses £4? It appears that the Special upright is losing money and that the Dome cleaner is the most profitable. These facts could lead to management making certain decisions. However, if we rearrange the figures on a marginal cost basis, a clearer perspective results:

Variable costs	*Cylinder*	*Special upright*	*Dome*
Direct wages	9	18	6
Direct materials	18	15	14
Direct expenses	9	5	5
Prime cost	36	38	35
Selling price	90	100	80
Contribution	£54	£62	£45

Selling one of each gives £161 contribution, which leaves a total overhead of £165, to give £4 loss as before. However, we can see that selling the special upright is more advantageous than selling the dome. This is the reverse of the previous analysis, which involved making decisions about overhead allocation before we made other decisions that are based on them.

It is better to ascertain cash flow towards the fixed expenses than to make premature decisions concerning profit potential on each individual sale.

Another distortion that occurs from using overhead allocation is from different activity levels producing different profit sales ratios. This is not the case for contribution/sales ratio.

The following example (Table 5.1) with simple numbers clearly demonstrates this effect:

Fixed cost = £5000, selling price = £10 each, variable cost = £5

Table 5.1 How profit and contribution rates are affected by the output. Numbers in brackets are negative.

Output 1	*Sales turnover* 2	*Fixed costs* 3	*Variable costs* 4	*Profit* 2−(3 + 4)	*Profit to sales* (%)	*Contrib.* (2 − 4)	*Contrib. to sales* (%)
800	8,000	5,000	4,000	(1,000)	(12.5)	4,000	50
1,000	10,000	5,000	5,000	0	0	5,000	50
1,200	12,000	5,000	6,000	1,000	8.2	6,000	50
1,400	14,000	5,000	7,000	2,000	14	7,000	50
1600	16,000	5,000	8,000	3,000	19	8,000	50
2,000	20,000	5,000	10,000	5,000	25	10,000	50
2,500	25,000	5,000	12,500	7,500	30	12,500	50

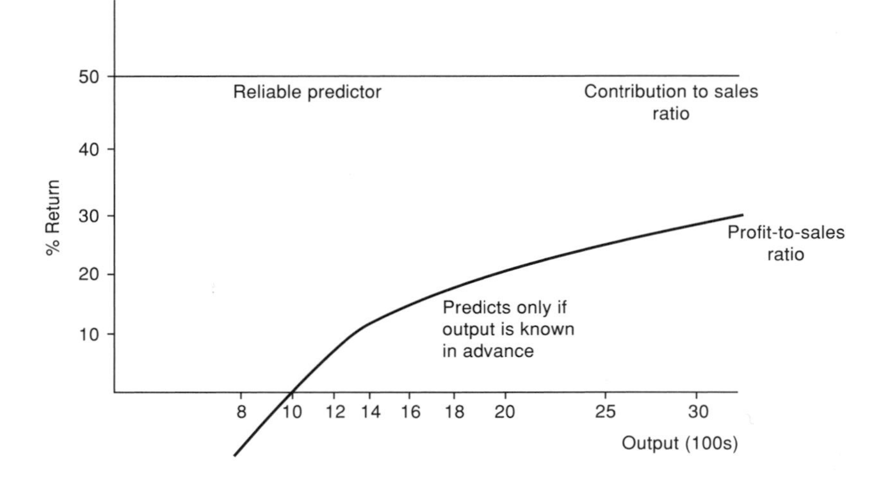

Figure 5.2 Differences between profit and contribution ratios.

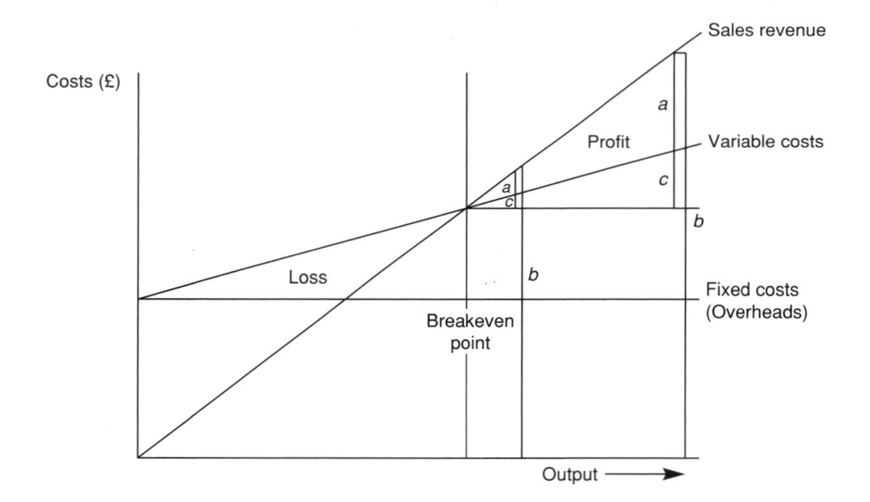

Figure 5.3 Breakeven chart showing marginal cost/contribution effectiveness

The results can be graphically demonstrated as a trend (Figure 5.2).

The contribution to sales ratio is useful in determining future profits when fixed costs are known. Obviously the profit-to-sales ratio can only slowly approach the set contribution-to-sales ratio. If the contribution-to-sales ratio is relatively low (say 20 per cent) then this holds down the result obtainable once overheads have been paid off. Hence, knowing the contributions made by each product signifies the strengths and weaknesses of the product range, in monetary terms.

A breakeven chart can be used to show the effectiveness of marginal costing, see Figure 5.3. By observing the slope of each line, certain deductions can be made.

When the breakeven point is reached, overheads are completely covered, and only sales and variable costs should be considered.

The ratio of a:b is profit to sales and this can be seen to be changeable, but the expression $b - \frac{c}{b}$ is consistent and the contribution to sales ratio

Why bother with allocated costing when once the breakeven point is reached, only sales and variable costs should be considered, that is marginal costing?

D. Introduction to the Problems of Waynes Scales Ltd

The company has been operating in the market with a standard range of electronic weighing equipment for many years, developing from a mechanical technology into an electronic one. The customers are mainly industrially and commercially based, and require a reliable high tech product from a proven market leader across the UK and Europe. This is what Waynes endeavour to provide.

The ideas and operating systems used for costing products had originated many years ago in the mechanical era and have been modified from time to time at the request of various managers and accountants.

Essentially the present system relies mainly on the one-time popular method of spreading the fixed costs on to the products using the 'percentage on labour cost' philosophy. Once again the validity of the methods used are being questioned. It is important to get the costing system right because costing is an aid to decision making, and incorrect or misleading information can do harm to the company's prospects. Remember, costing is only an aid and not the decision maker. More than money is involved in management. A costing system that misleads management is a double punishment because it provides the circumstances of paying for the privilege of being misled.

Questions about the system used arise because of one particular order which the costing system identified as losing money for the company. The manager of the department that fulfilled the order was convinced the order was efficiently and effectively completed to better target figures than other products that reportedly made good profits. He defied anyone to show him what extra could have been done to generate a profit. The present cost accountant, Samuel Wilde, had been at Waynes only 14 months and had his own worries about the costing system performance. He favoured some form of marginal costing but had received little encouragement when he reported this to the Managing Director at the time.

All these issues came to a head at a regular Production Meeting and the outcome, after a heated discussion, was to set up a task force to investigate different ways of costing the company products and also the sensitivity of profit margins to any proposed changes. The task force was to consist of Samuel Wilde, Joseph Gardner, Production Department Manager, and a consultant from HQ group services, later identified as Roy Jones, to advise on the group's costing tactics and experience. A number of products from the company range were to be analysed, chosen because they were considered significant, that is expensive and popular. Each would be subjected to different costing system analysis and a report with recommendations produced. It was expected that in ten days the data could be verified and the comparisons made in a further four days. The report with recommendations would be forwarded to the company Managing Director for further consideration prior to any action taking place.

E. Questions for the Reader and References

Possible Alternatives Assessed

During the data-collection phase the task force outlined the particular situations for important products being manufactured.

For brevity, only four such analyses have been included for the reader's attention. The problems have been structured to ease the learning process, so each should be attempted and conclusions drawn before looking at the information that the company considered and the actions they took.

Suggested references for further information

Costing, T. Lucey, 5th edn, DP Publications (1996).
Introduction to Cost and Management Accounting, R. Storey, Macmillan (1995).
Management Accounting for Decision-makers, G. Mott, Pitman Publishing (1991).

Four Case Studies

Question: Product case 1

A department is considered a cost centre and manufactures just two products, X5 and Y6. The task force obtained the following data and verified it as factual and repeatable on a week-by-week basis.

Cost centre fixed overhead is £8000 per week

	Product X5	*Product Y6*
Selling price	£10	£36
Material cost/unit	£ 2	£15
Labour cost/unit	£ 2	£ 5
Output per week	1000 units	200 units

(1) Determine the profit per week.
(2) Determine the profit from X5 and Y6 based on costing methods of:
   (a) percentage on labour overhead recovery,
   (b) percentage on material overhead recovery,
   (c) overhead recovery per unit.
(3) Compare and contrast the results obtained.
(4) Determine the sources of contribution towards overhead payment by products X5 and Y6.

Question: Product case 2

Waynes Scales Ltd makes plastic housings of a standard size. Each housing contains 3 kg of raw material which is moulded by machine, after which the finished article is checked and packed. The following information has been obtained from an investigation:

(a) Raw material is bought in 24 kg bags costing £30 each. Packing materials cost £0.25 per housing.
(b) The variable cost of machine time, including operator wages, amounts to £15 per hour.
(c) Two persons are employed on packing at an hourly rate of £5 each, and together take 2 minutes to pack one housing which is just keeping pace with production.
(d) The total fixed overheads for the housing department area amounts to £3000 for a 37-hour week.
(e) The selling price is £6.50.

Calculate:

(1) The cost of each housing on a marginal cost basis.
(2) How many housings must be made and sold to recover the fixed overheads fully.
(3) The new breakeven point if a raw material price rise of £10 per bag occurs.

Question: Product case 3

Waynes Scales Ltd manufactures electronic metering units and budgets to sell 36 000 a year, although the factory has the capacity to produce 40 000 units in normal circumstances. Variable costs are:

Wages	£2.00
Materials	£8.00
Overheads	£4.00

Fixed costs for the year are expected to be £201,600. The selling price is £20 per unit.

Calculate:

(a) The number of units to be sold to recover the fixed overheads fully.
(b) The budgeted profit for the year, assuming opening and closing stocks are the same.
(c) The number of units that must be produced to enable the amount of capital invested in production, that is £260,000, to return a 15 per cent yield.
This is the criterion for success specified by the Board of Directors.
(d) Explain the problems associated with the above answer.

Question: Product case 4

Waynes Scales has to decide whether to continue making a particular product or to buy it from a subcontractor. The selling price can be ignored if costs are concentrated on. A quote for £18 each has been received and as 50 000 per month are required, the product's removal from the schedule will ease the manufacturing and production control departments' work loads. Purchasing would increase clerical and storage/handling costs. These were included in an appraisal carried out by the cost department, consisting of the following:

Make or buy appraisal

Cost to manufacture		*Cost to purchase*		
Direct materials	£ 8.00	Purchase price	£18.00	
Direct labour	7.00	Storage	0.60	
Variable overheads	2.75	Clerical	0.03	
Fixed overheads	3.50			
Cost	£21.25		£19.63	Saving £1.62

50 000 × £1.62 = £81,000 savings per month

(a) Is the best procedure to make a 'make' or 'buy'?
(b) Is this analysis correct? Can another type of appraisal be produced? If so, compare the two methods.

F. Proposals Adopted and Company Solution

Case Study Facts Discovered and Analysed by the Task Force

The team discovered many similar instances to those included previously and took appropriate action. A marginal cost system was recommended and introduced to identify the high contributors towards the company overheads and profits effectively. Particularly useful, in the opinion of the Managing Director, was the data given to the salesmen concerning which products were the best earners. They were able to 'push' these products to the customers in preference to the low contributors. The whole policy of whether to make or buy was also analysed on this basis. Low contributing products, if they could not stand a price increase, were replaced by new higher contributors to enhance cash flow; while small subcontractors were used to supply the rejected low contributors.

The total exercise could be demonstrated by the problems encountered by a small department that made two products (Product case 1). The department had been used as a cost centre and overheads per week had been allocated. However, the problem concerned the question of what contribution did each product make to produce the total profit, which was known with some certainty. The following solution uses simplified coded data and demonstrates the previous costing information system for decision making and the newer marginal approach.

Solution: Product case 1

Two products labelled X5 and Y6.

Data identified: £8000 per week fixed overhead.

	Product X5	*Product Y6*
Selling price	£10	£36
Material/unit	£ 2	£15
Labour/unit	£ 2	£ 5
Output per week:	1000 units	200 units

The team analysed the profit per week and determined the importance of the roles of X5 and Y6 using a series of different systems. Then, finally, they utilised marginal costing to compare and contrast the results.

Profit per week

Sales = £10 × 1000 + £36 × 200
= £10,000 + £7,200 = £17,200

Expenses X5 = £4 × 1000 + Y6 = £20 × 200
= £4000 + £4000 = £8000

Profit = Sales − (Fixed costs + Variable costs) = £17,200 − £(8,000 + 8,000)
= **£1,200**

Now the £1200 had been identified, an explanation of its origin was needed to establish the relative importance of each of the outputs of X5 and Y6.

An explanation of profitability was first obtained by using the present system of a percentage added based upon labour content.

To calculate percentage on labour content, use

$$\frac{\text{Total overheads/week}}{\text{Total wages/week}} \times 100 = \frac{£8000 \times 100}{£2000 + 1000} = \frac{8000}{3000} \times 100$$
$$= 266 \text{ per cent}$$

Profitability of X5 + Y6

	Prod. X5	*Prod. X6*	
Sales	10,000	7,200	
Materials	2,000	3,000	
Labour	2,000	1,000	
Overheads on labour	5,300 +	2,670	= £8000 overhead allocation
Profit	670	530	
Profit/unit	67p	£2.65	

Note: Sell another 100 of product X5 and the company makes £67 profit!

The second method tested was the system of percentage added to material usage, and this gave the following results:

$$\frac{\text{Total overheads/week}}{\text{Total matl. usage/week}} \times 100 = \frac{£8000 \times 100}{£2000 + 3000} = \frac{8}{5} \times 100$$
$$= 160 \text{ per cent}$$

Profitability of X5 + Y6

	Prod. X5		*Prod. Y6*	
Sales	10,000		7,200	
Labour	2,000		1,000	
Material	2,000		3,000	
Overheads on material	200	+	4,800	= £8,000 overhead allocation
Profit	2,800		(1,600)	
Profit/unit	£2.8		(£8)	

Note: Sell another 100 of products X5 and the company makes £280 profit!

As products X5 and Y6 have some similarity, the third method tried was the overhead allocation to each unit produced:

Total units + 1000 = **1200**

$$\frac{\text{Total overheads/week}}{\text{Total units produced/week}} = \frac{£8000}{1200} = £6.6 \text{ per unit}$$

Profitability of X5 + Y6

	Prod. X5	*Prod. Y6*
Sales	10,000	7,200
Material	2,000	3,000
Labour	2,000	1,000
Overhead on units	6,600	1,400
Profit	(600)	1,800
Profit/unit	(60p)	£9

Note: Sell another 100 of product X5 and the company makes £60 less!

How costing information can be misused is demonstrated by the situation a 'trusting soul' encountered when driving home one night. He was listening to the news on his radio and the newscaster when reporting on a Shell Petroleum press release stated 'Shell have reported today that they lose 5p on every gallon of petrol sold and want to increase prices'.

The driver, reminded of his own need for petrol, drove past his usual Shell service station and called in at the Elf service station. He thought to himself 'Why make Shell lose 25p when I can make the French suffer instead'.

At this point the task force considered the vagaries of allocation had been exposed and company managers were more predisposed to finding a workable alternative. Finally, marginal costing was used to see if the results could be more meaningful.

Contributions of Product X5 and Product Y6

Marginal costing

	Prod. X5	*Prod. X6*
Sales	10,000	7,200
Labour	2,000	1,000
Material	2,000	3,000
Contribution	6,000	3,200 = £9,200 − Overheads
Contribution per unit	£6	£16 = **£1200 Profit**

Note: Sell another 100 of product X5 and the company makes £600 towards overheads and profit. As the overheads have been paid off previously, the £600 will swell the profits to £1800. This is a more reliable predictor of future events than the allocation ideas of the past examples ever gave.

With a whole series of analyses carried out, of which this one was typical, the company decided to invest in time, money and training to convert the costing system into the more responsive marginal costing method.

The cultural shock of these changes resulted in a new style of management decision making and greatly assisted the company's competitive position.

A report from the task force was circulated to all member companies with the challenge – 'Is your costing system telling you the truth?'

Summary

Similar product ranges or services are comparatively easy to cost and the logic of any method used can be demonstrated. However, when products are dissimilar, then the problem of overhead absorption is maximised. This is why it is better to use a method that does not attempt to spread overheads across these different products but leaves them to be paid for by a series of useful contributions. Overhead absorption can be used effectively but only when the situation requiring a costing model of this type has been evaluated and critically analysed for consistency of output. The scale of variations obtained with the examples given should be taken as a warning that errors in these areas are not small and therefore can be completely misleading. It is a good idea to challenge all figures quoted to verify their authenticity.

Solution: Product case 2

(1) *Marginal cost of housing*

Material, 3 kg of plastic at $\frac{£30}{24\text{ kg}} \times 3 = £3.75$

Packing material £0.25 $\frac{0.25}{£4.00}$

Machine/labour: Output is 2 min. each, 30 per hour

Cost of machine/hour = £15 $\therefore \frac{£15}{30} = 0.50$p each

Packing labour = £10/hour $\therefore \frac{£10}{30} = 0.33$p each

Total marginal cost = Material + Machine + Packing

= £4 + 0.50 + 0.33
= **£4.83**

(2) Contribution is £8.50 − £4.83 = £3.67 each.
Number of housings to recover £3000 is

$\frac{£3000}{£3.67}$ = 817 housings (which is about 27 h work)

(3) Material increases in price to £40 per bag. Material cost is now £5.00 instead of £3.75, so marginal cost increases to £6.08 and contribution reduces to £2.42:

New breakeven quantity becomes $\frac{£3000}{£2.42}$ = 1240 housings (which is over 41 hours work)

The material price increase needs an increase in selling price or sales volume if losses are not to be made.

Solution: Product case 3

Breakeven occurs when the total contribution equals the fixed costs.
Marginal cost of unit is addition of all variable costs.
Wages + Materials + Variable OH = £(2 + 8 + 4) = £14 variable cost.
Selling price is £20; contribution is £(20 − 14) = £6.

(a) Breakeven Quantity = $\frac{£201,600}{£6}$ = 33 600 units

(b) Budgeted profit based on 3600 output and sales leaves (36 000 − 33 600) = 2400 contributions; for profit margin, £6 × 2400 = £14,400.

Return on capital = $\frac{£14,400}{£260,000} \times 100$ = **5.5 per cent**

(c) Capital invested is £260,000 and 15 per cent return gives

$\frac{15}{100} \times 260\,000 = £39{,}000$

£39,000 profit to come from sales above 33 600.
Hence £39,000 = (X − 33 600) £6 X = Required output

$X = \frac{39\,600}{6} + 33\,600 = \underline{\mathbf{40\,100}}$

If all 40 000 maximum output was sold, only £38,400 profit is achieved which is under the target by 100 × £6 = **£600**.

Some part of the strategy needs changing so that targets set can be achieved.

Savings must be made if 15 per cent is to be achieved. Using maximum capacity is extremely difficult to manage on a regular basis because of various interruptions.

Can capacity be increased or the 15 per cent target be reduced?

Solution: Product case 4

Allocation confuses the situation because companies earn money for the overheads but in purchasing they spend. Use a marginal costing approach.

Second appraisal

Purchase cost as before £19.63
Manufacturing cost savings if purchased

Direct material	£ 8.00	
Direct labour	£ 7.00	
Variable overhead	£ 2.75	
Marginal costs saved	£17.75	Contribution £17.75 to existing Fixed overhead £ 1.88

Lost contribution: 50 000 × £1.88
= £94,000 per month

This high total would need to be apportioned to other products.

The method of marginal costing has shown the true effect of change; the response should be to transfer work out, but to choose those products with the smallest contribution. If this is the smallest contributor and most appropriate then go ahead, if not, keep searching for a negative contributor.

6 Operations Management of Plastics Extrusion

Section

A. Learning Outcomes

If the performance of a piece of equipment or department is below the performance targets set then something must be done. However, if the attempts to remedy the situation are unsuccessful, because of something lacking in the management's approach, then a state of equilibrium is often reached where poor performance is the accepted norm.

A more positive critical approach which will not accept the above is required, which is easy to say but more difficult to achieve when positions of people involved become entrenched.

This case study will demonstrate that:

1. A department with different work patterns, isolated from the rest of the factory, can be left to its own devices and be ignored by some managements, but to what result?
2. The isolationist culture forms bonds that become antagonistic to the needs of the firm, and the requirements of the department come first.
3. Targets never achieved are not targets at all; as more excuses are applied, the lower targets become accepted standards.
4. Lack of true knowledge concerning the activities in the extrusion department clearly shows that the management is not in control.
5. Operation control has to be worked for and monitored continuously, to be successful; otherwise failure is guaranteed.

B. Learning Objectives

After studying and analysing this case the reader should be able to:

1. Explain the importance of direct involvement and systematic analysis by a committed investigator/team.
2. Identify key features of how an investigation will succeed.
3. Understand why the problems of a specialist department become isolated if ignored, and result in a high cost to a company.
4. Differentiate between the shopfloor attitudes to the problems versus management's resigned attitude to lost control.
5. Define the reasons for the successful outcomes of the new management initiative.
6. Show that challenging the reasons for failure, and listening to staff monitoring the facts, lead to a series of simple solutions to what was seen as a complicated series of interconnecting processes.

C. The Basics of Operations Management

Introduction

Operations managers who control manufacturing units are responsible for the usage of the equipment, which will eventually change raw material into a sellable quality product, as well as the labour and capital inputs necessary for success. To bear this responsibility effectively they need to establish the methods, procedures and facilities that will achieve the targets set by the company. These component parts work successfully if the necessary detail they demand is effectively put in place because their relative importance is known. Many companies use Measures of Performance (MOP) and Productivity Measurements to monitor the ability of an operational unit to provide what is expected. Unfortunately, when measures are not appropriate to the standard set they cannot always provide reasons for any discrepancy. The people involved are then asked to do this and misleading excuses are not uncommon in these cases. This misinformation can form a large 'road block' to making the changes necessary for success.

Initially, each of the three main areas identified in the control of work is examined; each one is directly involved in this case study and on many other occasions also. They are:

1. Productivity assessment,
2. Performance measures, and
3. Productivity improvement.

Productivity Assessment

Only internal company indicators will be mentioned; the international productivity measurements used to compare countries and companies is outside the role of this particular case study.

Productivity is an often misunderstood term. It is not 'product output' which it is regularly confused with. Productivity is the efficiency of production, that is how well something is done. The classic formula for efficiency and hence productivity is:

$$\text{Efficiency} = \frac{\text{Output}}{\text{Input}}$$

All the problems involved revolve around the interpretation of what constitutes output and input. Many different measures are used, including money, weight of materials, area of materials, sales in units, value added, people, time, floor area, rejects etc.

If the data used for measures of output and input are widely drawn, then changes in the ratio are not easily attributable to specific areas for either good or bad reasons. On the other hand, if the measures data are detailed and exact, relating to one specific activity, then it is more costly to establish and monitor but the 'sign-posting' of the actual changes that occur can be easy.

Some companies' management may only have one or two production indices to reflect upon their performance, and these managements tend to be largely ignorant of reasons for local changes. However, the extreme alternative is to know too much about all the changes that naturally occur day to day and treat each one equally.

Limits of operational success need to be established, so the Pareto idea of major exceptions being targeted can be used. Crudely, this could mean only performance indications that vary by more than ± 10 per cent are reported on an Exceptions Document. Otherwise, as many as 250 different indicators can be available for consideration. World-class firms do have targets and measures of performance (benchmarks) to help them provide the regular consistent output demanded by their customers. These firms have a pyramid of performance indicators running from the top to the bottom of the organisation. They can be classified into three main areas:

(a) *Objectives* To help establish overall organisational effectiveness.
(b) *Strategic* To assist in providing benchmarks versus their competitors' results in order to highlight strengths and weaknesses.
(c) *Tactical/Operational* Detailed information based on shopfloor performance.

Generally productivity indices relate to costs, quality and delivery, and all three are central to a manufacturing area performance.

Examples of Performance Measures for Manufacturing

1. *Stock turnover ratio*, which is:
   Value of material used divided by current value of gross stock (cost)
2. *Sales value per employee*, which is:
   Total sales volume divided by the number of employees (cost)
   It can be used at many levels.
3. *Quality achievement*, which is:
   Number of items defective (scrap and rework) $\times$ 10^6, divided by the number of items processed (quality)
4. *Production achievement*, which is:
   Sum of items manufactured in a period $\times$ 100, divided by the number agreed on the production schedule (delivery)
5. *Manufacture lead time*, which is:

   Elapsed time between the order for components from a department being received and the components being dispatched (delivery)

   A further subdivision that is popular is to make productivity relate to the main inputs of a manufacturing department:

6. *Labour productivity*, reflects man-hours and wage rates compared with standards set.

7. *Material productivity*, reflects materials used and material costs compared with standards set.
8. *Plant productivity*, reflects capacity used and investment costs compared with plan.

Most measures are single-factor ratios but some multi-factor ratios are used when appropriate. They are called multi-factor because separate inputs are included, such as:

$$\text{Productivity} = \frac{\text{Output}}{\text{Input}} = \frac{\text{Output (£)}}{\text{Cost of labour, capital, energy materials and expenses}}$$

The classic multi-factor ratio is obviously

$$\frac{\text{Profit}}{\text{Sales}} \times 100$$

which includes all activities, but reasons for a change are also multi-factored.

Productivity measures are the life blood of successful change when dealing with a particular complicated process like the extrusion case study. They are reported but, if no action is taken, eventually the newer poorer performance becomes the standard and a downward spiral of effectiveness is set in motion. The interpretation fog which can ruin the usefulness and clarity of any productivity performance measures used can be demonstrated with an example.

Question

How would you measure the performance of a motor car? Which particular measures of performance would be best? Give reasons. Do not read on, but think and note your response.

The answers obtained to this question are often different and wide ranging.

The following quoted answers are useful in explaining the inherent difficulties of putting indicative numbers on to achievement.

1. Top speed,
2. Acceleration time to 60 mph,
3. Drag factor,
4. Miles per gallon (general),
5. Miles per gallon at 56 mph,
6. Cost divided by useful life,
7. Total miles that engine can achieve.
8. Brake horse power available,
9. Time to reach 60 mph and stop,
10. The amount of calorific value of fuel that is changed into movement,

11. Cost per mile to run over x years,
12. Relative comfort of ride (suspension),
13. Cornering ability,
14. Safety features in collisions,
15. Resale value,
16. Cost of servicing and regularity,
17. Body resistance to corrosion,

plus many others based on previous indicators.

Each one of these indicators can be useful in some way but, if all of them are used, how can a reasonable judgement be made between cars? The conflicting nature of some indicators, some being better and some worse for different cars, means they only marginally help a comparison. With so many different measurements potentially available, the operations manager must be satisfied that the ones in use are satisfactory for the purpose and assist in achieving the company objectives.

Productivity Rules

1. Productivity indices are meaningful if:
   (a) Input and output can be unambiguously identified and defined,
   (b) The numerical values are based on consistent and reliable data.
2. Productivity indices are useful if:
   (a) People who use them understand how they are built up and the factors that affect them,
   (b) People can influence these factors, or a significant proportion of them,
   (c) People are supplied with information to support effective action by management,
   (d) The above is done promptly and frequently,
   (e) They are acceptable as a norm of productivity, that is they are not in dispute.

Indicators can be established and traditional, but then become outmoded when new methods, technologies and products are introduced; productivity measures then need re-evaluating to test these relevance.

Productivity Improvement Methods (PIM)

There are *three* basic means, at different levels, by which productivity gains can be made:

A. Research into new materials and processes: this is a long-term (5-year) aim which is expensive but also potentially provides the best productivity improvements.

B. Technical applications of new scientific knowledge to materials and processes: these can be medium term (2 to 3 years), are costly but provide good savings.
C. Operational changes to develop procedures, invest in people and eliminate as much delay and cost as possible.

The advantages of this final approach are that it is not expensive, savings can be effected in the short term (in days) and it requires the least capital outlay of all. It is by this means that improvements are primarily obtained in the following case study.

All three methods (A, B and C) are useful and can be used as a set, however poor performance and excessive delays mean large savings are possible in industry once the cloud of misleading information is penetrated. Japanese industry has effectively used Method Study (name changing is popular) Kaizen, to produce productivity gains of such proportions, compared with the West in chosen areas of expertise, that market domination is achieved on price and quality – not bad for a country with few natural resources.

When the savings using C are potentially smaller than these possible with the two technical approaches, its strength is continuous improvement. The first two, A and B, need to be followed by an effective operational style so that all the potential gains are achieved. Style C involves thinking the firm is in a race without a finish and improvement is the life blood of survival. Companies fail because of extra costs needlessly incurred by delays which starve them of money.

Systematic Approach to Difficulties

This particular case study on extrusion involved the investigators using an analytical, systematic procedure. One of the branches of Work Study is method study which has been defined as:

> *'The systematic examination of activities in order to improve the effective use of human and other material resources'*

It is systematic because it follows a step-by-step procedure which does not allow anything to be overlooked. Also it is analytical because it is based on facts established and not on the opinion of people involved, who may have vested interests.

While improvements can be made as a result of insight, doing the obvious or 'trial-and-error', applying the method study approach makes savings more certain and more secure. This ability comes from the following:

1. Looking at the activity as a whole, not piecemeal.
2. Sometimes minor changes are easy to identify but a change strategy is best validated by a close investigation.

3. Alternatives are analysed, making choice more reliable.
4. Techniques and experience can be applied once the critical analysis has been carried out.

Method Study: Sequential Steps (see also Chapter 9)

Here is not the place to write a book on Method Study, there is only space to outline the advantages and reasons for adoption (see references given later). The essential characteristics of this methodology are used in this case study but modified to suit the prevailing circumstances. The overriding philosophy of adaptation is the key to success because a robotic application of the steps involved will not produce the desired effects. This systematic procedure, to be applied to any existing task, is as follows:

1. Select the work to be studied

Know the reasons why the investigation is taking place and the changes required in output to achieve them. Without a clear view of the output part of efficiency, the input cannot be reduced and matched to resources.

2. Record the present method

Here the facts are established concerning the present system, not the opinions and negative thoughts of the unenthusiastic. Charts, diagrams and symbols are used to condense all the activity into a form that can be analysed. The fewer words used the better, and special emphasis is placed on identifying delay and waste as an enemy of performance achievement.

3. Examine every part of the present method critically

A series of well-known questions are posed to elicit the potential for advantageous change. The questions, sometimes known as (1H, 5W), are to be answered by the investigator after fact-finding, they are not questions to put before anyone else involved in the study. A simple matrix sheet of questions and answers assists in condensing the facts and placing them against the analysis and challenges of: why it is necessary? Possibilities available and chosen recommendation can then be justified. (1H, 5W) is short for How?, Where?, When?, What?, Who?, Why?.

4. Develop a better method of achieving the task

As the scope for change has already been established in step 3, this base can be used to develop methods for effective change. The investigators' experience, techniques of change (standard basic solutions) and advice from consultation can all be brought to bear on the problem.

5. Discuss and agree the recommendation

This is an evaluation phase to bring all involved into a position of support. The recommendation will be the best solution, considering the prevailing circumstances, to ensure the full support of all. Signed pro-formas can formally establish this is indeed the case. People will not lightly sign to say they are satisfied if they are not.

6. Train those involved in the new method and skills required

Preparation for success is essential if the investigators long-term credibility is not to be compromised.

7. Install the new method

Each stage should be planned, based on feasibility, with a nominated person in overall control. No benefit is obtained until this step is completed. Again a signed pro-forma is the only acceptable means to establish that the installation is complete, that is by agreement. Any problems will manifest themselves at installation, so they must not be ignored or down-graded if the culture of successful change is to be encouraged.

8. Maintain the new method to retain the benefits

Regular checks and responses to information obtained are needed because all the usual stresses of operational control may lead to an inferior alternative being used in place of the one required (see the reference list for further reading on the methods necessary).

D. Introduction to Problems at Heartfords

Heartfords Ltd is a company that makes a range of flood and fluorescent light fittings for the UK and European market. Its track record had been good but, over the past few years, the effects of foreign competition, in a poor market, have had an effect. In response, a new product has been designed and tested as a prototype. However, projections and forecasts suggest that the overall cost may be too high for the volume planned. The cost of all processes, as well as design features, are under scrutiny to identify savings and efficiency changes.

One of the major processes used in the production of the light fittings is extrusion of the diffuser. This case study concentrates on this particular problem area and the changes needed to reduce the costs of production. Naturally other areas of production were tackled in a similar way to make the fitting eventually very competitive.

The range of diffusers in question is only one of many styles and types produced in the extrusion department, for example variations are based on different lengths, cross-section profiles and material. It is believed that the better performance planned for the new diffuser will result in better performance overall. This is the background that started the investigation into a production area that had been neglected by management for many reasons for many years.

General Note

In conditions where regular changes in methods and procedures occur to improve efficiency, savings are modest but welcome. However, in areas where little has been done over a period of five, ten or fifteen years the savings are spectacular and headline seeking. Unfortunately this clearly shows how much money has been wasted/lost by the company operating at low efficiency levels. This is the trap that progressive companies do not fall into.

The skill of sorting the important facts from misleading opinions, is required when obtaining data from an operational department, as is the ability to determine that data are missing from the records either by omission or commission by certain interested parties. The politics of vested interest are rarely absent from this activity. Newly qualified engineers cannot simply read about theories and systematic practices and then successfully carry them out. This skill has to be nurtured and developed progressively elsewhere, but at least the inexperienced investigative engineer needs to be aware of the difficulties. With this in mind, the following data outline the information eventually gleaned regarding this one development. Generally it is important to look at the total needs of a company so that improvements planned are not localised, resulting in company suboptimisation. In this particular situation a sharp focus of waste and inefficiency was identified. It was realised that generous benefits for the company could be obtained, albeit with some effort on behalf of the investigator.

The Present Position Outlined

A change in management had caused some of the areas of inefficiency previously mentioned to be readdressed. The indicators used in the extrusion department only gave the 'tip of the iceberg' information according to the new General Manager. On his regular visits he found that scrap and rework were at extremely high levels; this attitude was based mainly on observation. The supervisors and workers in the department could not explain these apparent losses and did not seem much interested either. Action was clearly called for, so the Group Industrial Services were contacted and a leading consultant, Joanne Walters, was scheduled to make recommendations within two months.

She formed a team of three, herself along with a newly qualified manufacturing engineer and the extrusion department supervisor. Each was allocated an area of activity in the investigative plan. The graduate trainee was to work full time in the extrusion department tackling most of the jobs. This would allow him to monitor the operational efficiency by making actual measurements of activities rather than letting the official company documentation tell a different story. The areas of particular concern were quickly identified and monitored, and consisted of:

- Changeover time from one extrusion profile to the next.
- Time elapsed between closure of one product and production of the next to specification, that is to good quality.
- Weight of material scrapped, reground, and used again (three totally different quantities).
- Reasons for scrap, breakdowns, absenteeism and tooling difficulties.
- Manning levels during the three-shift 168-hour pattern of working.
- Work practices that the skilled setters used to change from one extruded product to another.

The graduate was supported in his investigative style by the opportunity to call on the supervisor for comments and reasons for particular actions observed. It is more difficult for a supervisor to 'fob off' a witness to problems, especially when it is known that he will still be watching proceedings for weeks. The team met regularly and planned a strategy of continuous improvement, tackling one identified 'bottleneck' after another on the interconnected continuous process.

One of the reports produced by the team is summarised as follows into facts discovered and problems identified to enable the overall picture of inefficiency to be revealed.

The main productivity indices used for the extrusion department related to workers' performance against a set time standard and sales value of production achieved per month. No data were taken or analysis conducted on scrap produced, equipment usage, power or reclamation.

Facts Discovered

- Six plastic extrusion machines were available: one concentrates on sheet production and five on diffusers.
- The working pattern was 24 hours a day (three shifts of 8 hours), seven days per week.
- The plant valuation with accessories and tools was £1.5 million.
- There were twelve workers on each shift (two setters and ten operators).
- The average time elapsed between production runs was 3 to 6 hours; some, however, were in excess of 8 hours (1.5 hours was the company standard).
- Observations showed that more than three extrusion machines were rarely working at any time, and sometimes only two.

- Scrap rates were in excess of 10 per cent, and were largely ignored by everyone because it was thought that defective plastic could be reground and recycled. (This was found not to be true – much plastic was not recycled since its condition was poor, that is 'burnt', so it was put into a tip.)
- Large quantities of work in progress, scrap, material waiting for regrind plus the new material in containers lay around the whole shopfloor. This was the main contribution to the poor 'housekeeping', the department looking an unsightly mess with material and tools spread around chaotically.
- The production schedule achieved was never more than 75 per cent, which meant that only 75 per cent of the necessary diffusers for assembly and customers' requirements during the month were made, in spite of apparent spare capacity (machines standing idle).
- Only 70 per cent of material purchased found its way into finished saleable products, even with the advantage of regrind/recycle potential of some plastics. The rest was scrapped, with no cash return, because of contamination or burning/overheating.

Problems Identified

It should be noted that the following problems were found by observation and questioning, because not all the information required was recorded on either documents or computer files by the company

- Dies for extrusion of diffuser profiles were not aligned with the extruder head, so time is lost while this is achieved during the start-up run, producing scrap until all is correctly in place.
- None of the process equipment was fixed or aligned with the extruder (see Figure 6.1); each can be bolted to unspecified floor positions.
- 'Burning' (darkening) of the plastic material occurred in the screw feed, not from heat control problems but because of the low extrusion speeds used. Hence, material was heated for too long in the extruder barrel and screw.
- The cooling system used for the extruded material was by cool air jets passing on to the surface of the plastic. Air jet positions varied in configuration for each extrusion (see Figure 6.2a) as did pressure applied and volume of air used, which resulted in distortions to the extruded profile output. It was understood that irregular cooling gave irregular profiles but nothing in particular had been done about it. The operators tried different flows until an acceptable product was made.
- The airways designed to cool the plastic did not aid flow (see Figure 6.3(a)).
- The puller conveyor process equipment had damaged conveyor belts and needed some refurbishment.
- The automatic cutting machine had a variable-length setting system, which had proven unreliable, and slippage occurred over the calibration wheel

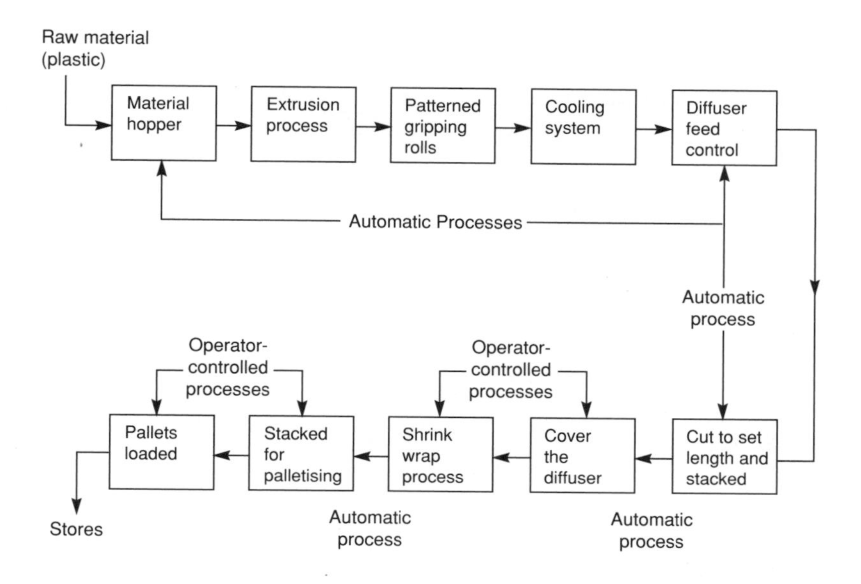

Figure 6.1 Processes involved in the extrusion of diffusers. The process positions are not in a fixed relationship.

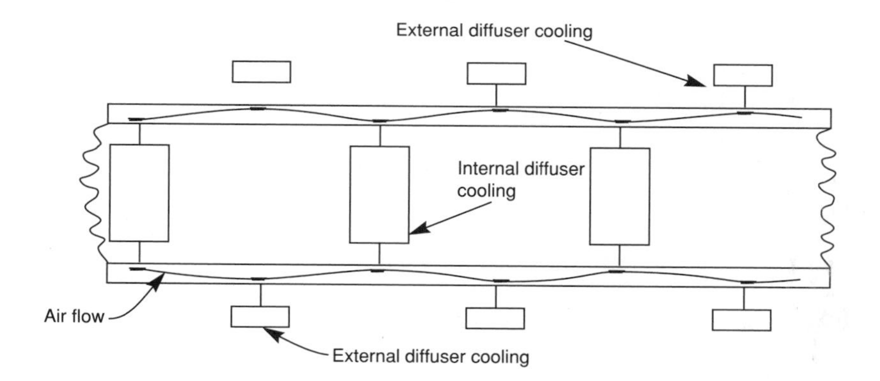

Figure 6.2a Wave formation (distortions) resulting from poor cooling

that measures diffuser movement, resulting in scrap and rework for diffusers, as they are then too long or too short.

- The shrink wrap machines were variable in operation, resulting in poor packing appearance which often meant removal and recycle. See Figure 6.4 for dual usage.
- The statistical process control charts that are used were either incorrectly filled in or ignored. Very little useful information could be gleaned from them, even when problems are rife.

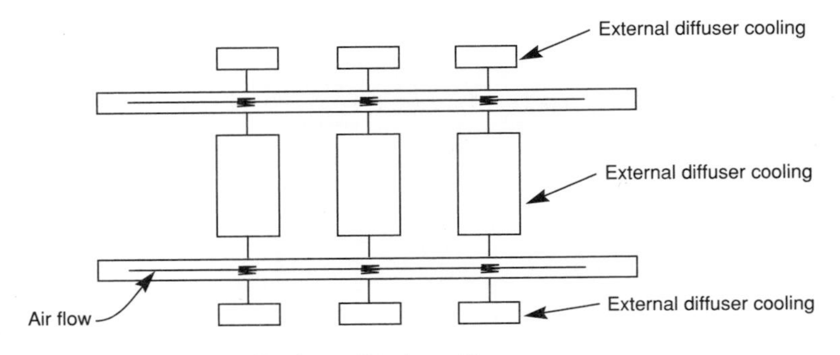

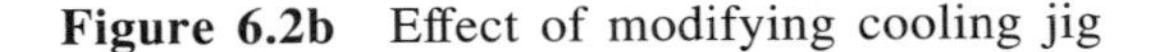

Figure 6.2b Effect of modifying cooling jig

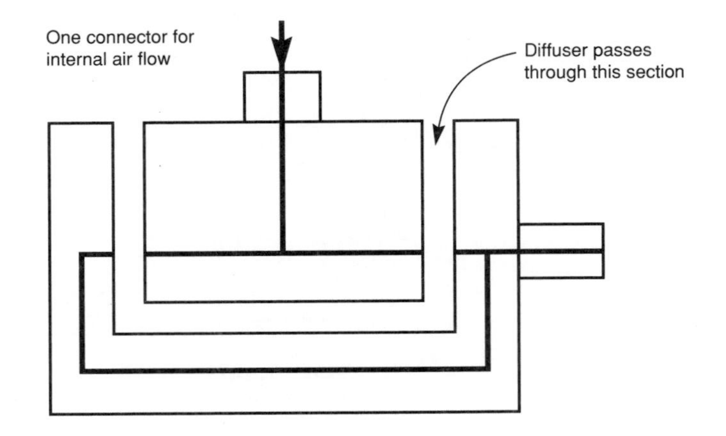

Figure 6.3 (a) Variable gap causes variable flow and cooling

- The workforce had been ignored for a long time by the management in as far as method and equipment changes are concerned. The only real contact point was through urging them to meet the schedule set down but which was likely to be changed during the month.
- Setters were employed for specialised setting and did no other activity. The rest of the team was flexible as far as different job operations were concerned.
- Initially some of the management thought that the extrusion department was satisfactory, with no need for radical change. Others thought £1000s per week could be saved with the right management input.
- Material for regrind and recycle was left lying around the department for so long that it became contaminated with dust from other materials, which meant it was unusable and worthless.

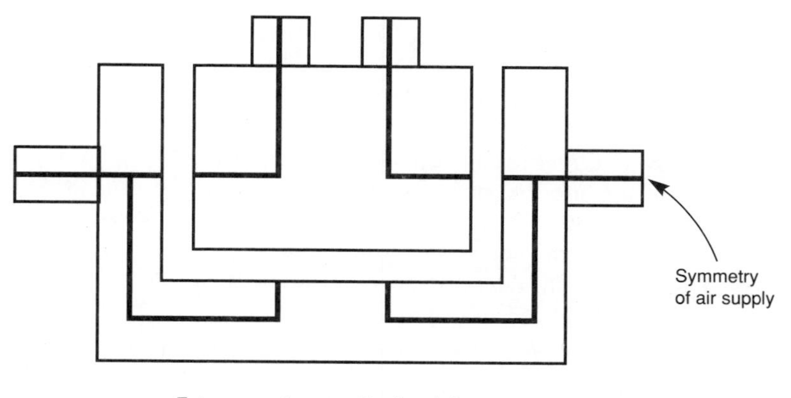

Figure 6.3 (b) Greater air supply gave faster cooling and speed of extrusion was increased.

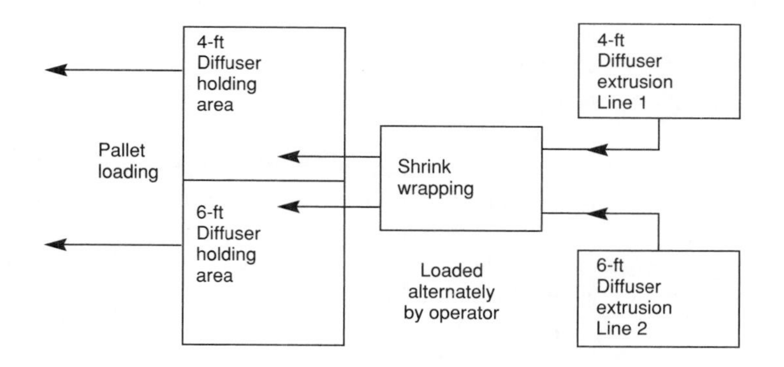

Figure 6.4 New layout for shrink wrapping machine to enable high utilisation of available capacity

Student analysis

It is not necessary to be an expert in plastic extrusion to see that changes are called for and where they are needed. This position comes from the list above, which spells out the core problems. The above investigative activity that produced the facts as stated is the hardest and most important part of the systematic methodology used. They provide the base on which sensible productive changes can be made.

E. Questions for Reader and References

Consider the Possibilities Available

1. What is the best way to tackle the problems identified?
2. Which problems are the most important and should be tackled first?
3. The interactive nature of each part of the process needs some attention. How is this best achieved?
4. How much time and capital expenditure may be needed to ensure that company targets are met?
5. Which cost savings can be set against the above question?

Remember, specific answers do not exist in text books to enable these particular problems to be solved. They provide the basic outline theory and insights into possible action. It is the application of these ideas and the experienced gained in developing the new ideas that contribute to company success. Some say it is 30 per cent theory and 70 per cent application based upon circumstances that provide the ingredients of improved performance.

Suggested references for further reading

Management for Productivity, J.R. Schermerhorn, 4th edn, J. Wiley & Sons (1994).
Manufacturing Systems, edited by V. Bignall *et al.*, Blackwell (1985).
Operation Management – Concepts, Methods and Strategies, M. Vonderembse and G. White, 2nd edn, West, USA (1990).
Production and Operations Management, D.M. Fogarty, T.R. Hoffman and P.W. Stonebraker, South Western, USA (1989).
Production and Operations Management, A.P. Muhlemann, J.S. Oakland and K. Lockyer, 6th edn, Pitman Publishing (1992).
Work Study, R.M. Currie, edited by J.E. Faraday, 4th edn, Pitman Publishing (1977).

F. Proposals Adopted by Heartfords Management

Joanne Walters' skills were demonstrated in the way each idea for change was presented and sold to the management and workers. She used facts that neither camp could deny to establish the need for change, and then chose easily 'winning' situations to build up her credibility with all concerned. By the time three new ideas had been well received and had shown improvements, the resistance to any further changes was non-existent. The degree of change and success will be demonstrated via the results that management saw and welcomed.

All the problems identified are important in an interactive continuous process system but some have to be classified as priority issues based on the above criteria. The first of these was clearly considered by Joanne to be the cooling jigs. The team realised that altering the flow of air through the orifices on to the hot plastic surface would have a dramatic effect on the whole process. Equal consistent cooling was achieved by moving the plates into alignment (see Figure 6.2b), this made cooling twice as effective, by removing the distorting source from the pliable plastic.. At the same time, flow ways through the plates regulating flow were changed (see Figures 6.3(a) and 6.4(b), assisted by reliable new pressure gauges for accurate pre-setting of the now recorded successful conditions for air flow. The increase in cooling efficiency obtained enabled the extrusion process speed to be increased by 40 per cent. Later this improvement was reduced a little (to 33 per cent) to balance other activity times on the diffuser-process production line. As a result of this speed increase, scrap due to 'burning' marks was eliminated ('burning' is primarily caused by slow speeds through the screw barrel).

The extra output obtained meant that the next part of the process investigated needed to be the shrink wrapping processes and their unreliability. It had been discovered that although four shrink wrappers were available, they were not highly utilised. Further, it was found that two were more reliable than the other two. It was considered that one shrink wrapper could serve two extrusion lines if correctly relocated to avoid lost time from operator movement (Figure 6.4), hence the two chosen machines could cater for all the output of the diffusers to be shrink wrapped. Some maintenance work was carried out at the same time to improve reliable performance. It was also decided that a longer-term solution was required for planned future expansion, so Joanne's team made plans to purchase two higher-output, refurbished models for delivery in six months' time.

The next investigation brought to a conclusion by the team concerned the 'cutting to length equipment', which was still producing scrap, mainly by cutting short or making rework by cutting too long on many occasions. New electric controllers were fitted using a fast setting method so that standard lengths of 6 foot, 5 foot, 3 foot etc. could be quickly reset by the operator. The improvement from this was most distinctive; scrap and rework were almost eliminated from this stage. The changes carried out so far resulted in a considerable increase in effective output which enabled the whole of the monthly schedule to be easily met, however the setting time between different product runs still dominated lost time.

Setting Time Reduction Changes

Joanne, with the help of the supervisor, produced a comprehensive plan to simplify each stage of the changeover routine. A detailed list of actions required and completed contained some of the following features:

1. Diffuser dies were modified to include slots and alignment setting pins were put into the extruder head so that different dies could be exchanged in minutes. The pattern rollers were fixed to rails embedded in the floor to ensure quick, reliable alignment with the die profile. The old 'floating' system meant that time was wasted in ensuring the pattern rollers were running square over the extruded section. Figure 6.1 shows the interconnecting processes. Pattern rollers give the distinctive diffuser pattern, whether straight lines or cross-hatch.
2. The condition of the gripper feeder was improved by renovation to exert consistent pull and grip, which helped reduce the lead time required from initial start-up through to producing a good product to specification.
3. Repeated set-up statistics were collected and revealed that, from stopping manufacture of a profile to acceptable production for a new profile, the time was reduced from the variable 2 to 8 hours to a consistent 30 minutes. This made scheduling predictions of output much easier and saved over 60 hours of ineffective time per week across the department's extruder capacity.
4. The combination of all the above changes had a dramatic effect, including reductions in scrap, rework and wasted time. The higher utilisation of all extruders greatly increased capacity. Scrap was reduced by £25,000 per year and rework by £35,000 per year.
5. A month's demand from the customers could now be produced in two weeks. It naturally followed that, after the workforce had been consulted, it was moved to a 'bottleneck' operation on an assembly bench elsewhere in the company for the other two weeks to aid overall company performance.
6. In addition to these savings mentioned, even more accrued from heating, lighting and power, because of the two weeks of inactivity. Preventative maintenance, which had been ignored when full activity was the rule, could now be applied during the shut-downs. Scheduling was now reliable and was further helped by employing certain different scheduling rules, such as the nomination of each extruder to a specific range of products. This meant that particular extrusion machines were scheduled to produce the needed diffusers and the resulting local specialisation assisted performance. Production operatives moved between extrusion lines as the customer demand required.
7. The final area to be addressed in a formal way by Joanne's team was the workforce. Obviously it had been directly involved during the changes but now further developments were needed to move towards 'world class manufacture'. The workforce was trained and organised into Continuous Improvement Teams to gain further savings by suggestions and discussions arising from their direct involvement. The first two meetings produced over thirty suggestions, all of which would be investigated to see how much they could aid productivity. At the same time, work normally subcontracted out to suppliers could be brought back inside the company to utilise the spare capacity now created. Needless to say, all these changes also had an affect on 'housekeeping'. Less scrap produced meant less

clutter, each extrusion line was painted in a bright colour, and interest and pride returned. Success brings increased status when all concerned have made a contribution.

Summary

This case demonstrates the extent to which situations can deteriorate when regular attention is not paid to identify and remove the problems that exist. The long list of shortcomings quoted shows a condition of inactivity on behalf of management, along with some disagreements about what should be done. Generally, in these situations it can be the workforce's 'apparent failure' that receives management's attention – the workers become the scapegoats. When a list of problems is identified and tackled in a systematic manner, the savings accrue from reduction in waste and delay. Waste and delay are the two enemies that engineers should always be trying to identify and remove. This is easy to state but takes commitment to realise useful results. The graduate engineer was made the manager of the extrusion department as part of his professional development and also to provide the input to ensure maintenance of all the gains made. It was considered that the old ways would return without the continuance of the 'outsider' influence.

Some of the final performance figures for the first year of operation were:

(a) Previous value of products produced per month was approximately £8000; this was changed to £12,000 using only two weeks' production capacity.
(b) Savings in scrap, rework, energy and enabling workforce to transfer to other work in the firm is in the order of £200,000 per year.
(c) Capital expenditure required to facilitate changes was only £6000 plus the later purchase of two refurbished shrink wrappers.
(d) The payback period for the capital expenditure was two weeks. The commitment to change demonstrated in a management team is the criterion necessary for action and results. Generating savings is a continuous process and the responsibility of management; a policy of leave alone is often wrong. If the management were to sit back and take no action from now on, they will eventually reproduce the situation that originally existed in this case.

Footnote

The new General Manager was curious why all these advantages and savings obtained, which he did not dispute, failed to affect any of the company performance indicators, especially in the areas of financial gain. He was forced to assume that elsewhere in the company similar delays were cancelling the advantage. Could the performance figures give him any help? That would be another case study.

7 Assembly Line Replanning

Section

A. Learning Outcomes

Material flow both to and through an assembly line is of major cost importance. Too much material (work in progress) increases delay, losses and costs. Too little material causes delays in production and efforts to correct the problems in a panic. To obtain the correct balance of costs and efficiency is an important aim of application engineers. Analysing the faults of a present system is a critical step towards achieving the desired results.

This case study will demonstrate that:

1. A considerable difference can exist between planned activity and actual activity.
2. The underlying problems can be hidden by an accepted lack of success and failed symptom correction.
3. Systems viewed in isolation, such as the kitting of parts for assembly,

can cause bottlenecks away from production that are just as effective at causing delays as any other factor.
4. Accepting the *status quo* for order quantities, without questioning their validity, results in rigid responses to the need for change.
5. Traditional methods of changing from one product assembly to another, in which overworked supervisors play the main role, are the cause of lost production, including a culture of rest periods between activities.
6. Simple methods that rely on the visual appraisal of the assembly situation are the aim of attaining material flow balance.

B. Learning Objectives

After studying and analysing this case study the reader should be able to:

1. Explain the usefulness of Just in Time applications to material movement.
2. Identify key activities in the determination of material and time wastage.
3. Define the parameters of visual control.
4. Identify good methods in areas of change.
5. Differentiate between simultaneous and sequential activities.
6. Identify the 'road blocks' put in the way of change.

C. The Basics of Improving Throughput Efficiency

Although the aim of this case study is to show an application of the 'Just in Time' philosophy, it is wrong to presume that this is the one and only problem-solving technique available. Other acceptable methods are available, as discussed below.

1. The capital-intensive approach

- Moves towards single operation manufacture.
- Factory-wide application of above.
- Uses sophisticated scheduling to satisfy customers.
- High capital investment is required, updated regularly.
- Problem exists of whether the capacity created is flexible or dedicated.

2. The simplification approach

- Breaks production into cells.
- Maximises labour flexibility.
- Monitors with process analysis.
- Arranges for visual control to dominate.

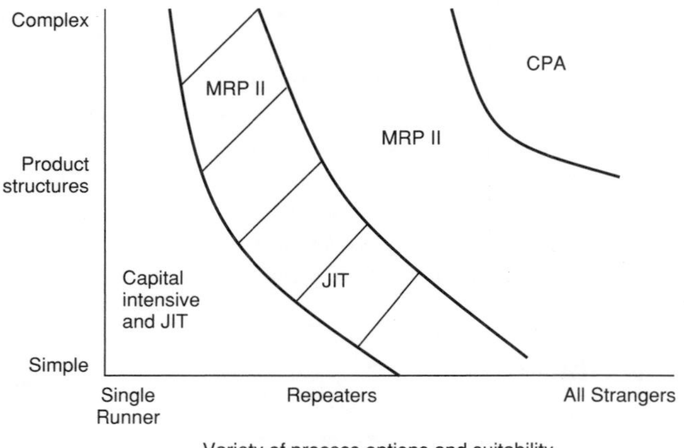

Figure 7.1 Just-In-Time application areas

3. The system-intensive approach

- Uses MRP II (comprehensive software package) for Manufacturing Resource Planning.
- Sophisticated data processing.
- Detailed scheduling.
- Useful in a variety of operations/broad applicability.

4. The Just-in-Time approach (Kanban and its derivatives)

- Low level of queues.
- Small transfer batches.
- Minimised set-up times.
- Good manufacturing engineering.
- MRP (Material Requirements Planning).

The different applications can be approximately positioned by consulting Figure 7.1, which simply demonstrates the areas of influence of each approach.

Basics of Just in Time

JIT is a philosophy that encompasses various concepts which can result in a different way of operating for most organisations. The basic points of the philosophy include:

1. All waste, anything that does not add value to the product service, should be eliminated. Value is anything that increases the usefulness of the product or service to the customer or reduces the cost to the customer.

2. JIT is a never-ending journey, achieving rewarding steps and milestones.
3. Inventory is a waste. It covers up problems that should be solved rather than concealed. Waste can gradually be eliminated by removing small amounts of inventory from the system, correcting any problems that ensue, and then removing more inventory.
4. The customers definitions of quality with their criteria for evaluating the product should drive product design and the manufacturing system. This implies a trend to increasingly customised products.
5. Manufacturing flexibility, including quick response to delivery requests, design changes and quality changes, is essential to maintain high quality and low cost, often with an increasingly differentiated product line.
6. The employee who performs a task is often the best source for suggested improvements in the operation. It is important to employ the workers' brains and not just their hands.

Traditionally inventory has been viewed as an asset, one that can be converted into cash. The JIT view is that inventory does not add value but instead incurs costs and therefore it is considered a waste. Again, traditionally, holding inventory was seen as being less costly than correcting the production and distribution inefficiencies that inventory can overcome. For example, large lot sizes spread the cost of expensive set-ups across many parts. JIT users take a different view and question expensive set-ups.

JIT views inventory as a symptom of inadequate management, a method of hiding inefficiencies and problems as shown in Figure 7.2. Inefficiencies that cause inventory include long and costly set-ups, scrap, lengthy and widely varying manufacturing lead times, long queues at work centres, inadequate capacity, machine failure, lack of worker and equipment flexibility, variations in employee output rate, long supplier lead times and erratic supplier quality.

Using the JIT method emphasises the importance of solving each of these problems which will reduce the need for inventory and improve productivity. The planning strives to have the right material at the right time, at the right place and in the exact amount. Therefore it can be said the name 'Just-In-Time' is used by many to designate an organised and continuing programme to improve operations productivity.

The main points

To restate in main point form, the JIT approach includes:

(a) Reduction of set-up times to achieve smaller production lot sizes.
(b) Increased use of sequential flow processes, such as dedicated assembly lines and group technology cells.
(c) Increased use of multi-function workers.
(d) Increased flexibility of equipment and capacity.

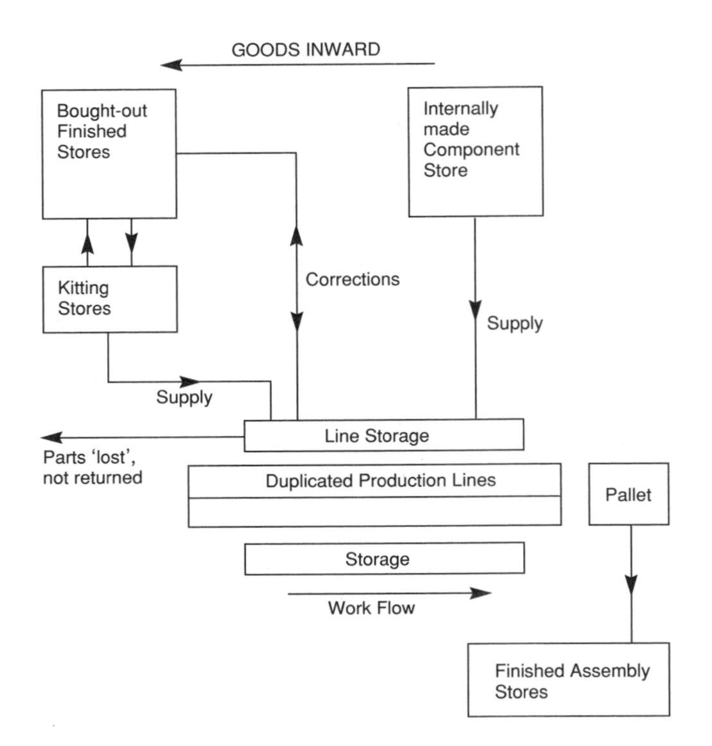

Figure 7.2 Poor materials management before JIT installed

(e) Increased use of preventive maintenance.
(f) Increased stability and consistency in the schedule.
(g) Long-term relationships with suppliers.
(h) More frequent deliveries in smaller batches from suppliers.
(i) Improved technical support of suppliers.
(j) Employee involvement programmes, such as quality circles.
(k) Statistical process control.
(l) Cause and effect analysis.
(m) Delivery direct to work area, avoiding double-handling into and out of stores.

Work in progress is the easiest symptom to identify. Problems of work in progress are many and have the effects of queues, missed delivery dates, costs of storage and lack of overall control (see Chapter 4). Traditionally, production is 'pushed' through each process by a demand created at the first stage of a sequence of operations, which is then transferred via queues to subsequent processes. Eventually the product clears all stages and advances into the finished parts stores. It is this 'push' procedure, often accompanied by progress-chasers to encourage movement of particular parts at the expense of others, that appears to be dynamic but it is basically flawed.

For assembly operations, as used in this case study, one way of preventing excessive stock or work in progress is to limit the material flow to the assembly line by making sets of parts to enable the required quantity to be built and no more. This process is called *kitting*. In theory, kits of parts, just enough to fulfil an order, seem a slimline way of achieving satisfactory results. However, it actually adds another process (non-value-added) to those already required that is time-consuming and subject to delays and errors in the quantities provided. Each of these problems then manifests itself during assembly activity, causing delay and increasing cost. The kitting activity does not alter the push philosophy of the production process.

This is where the opposite philosophy of JIT can show advantages. The idea is based on 'pull' rather than 'push'; that is, the movement of work is related to the demand on the finished part stores, not the order requirement being fed into the first stages. Coupled with this, restrictions are set at each stage concerning the quantity allowed to be stored by any process.

This effectively reduces work in progress to a set, minimum figure and the only movement allowed is when space is provided by goods moving towards the stores. This control is obtained by providing a set number of component containers for a system which restricts the capacity for storage. The system is visual and could be said to have something in common with kitting, but it is minimalist and accurate because shortage problems can be seen early and action taken with far fewer resources used.

Figure 7.3 show the two philosophies of 'push' and 'pull' alongside one another.

The work movement is typified by a 'rope' which while in tension, that is, 'pull', will move as it is required by the demand. However, when subjected to a 'push' system (which has a small element of pull because the stores wants parts and progress-chasers affect the process), then inevitably the rope moves and creates 'windings' which are areas of extra rope (work in progress). These disconnections prevent the rope from functioning and represent loss of management control. As mentioned before in Chapter 4, the concept of runners, repeaters and strangers still operates in this area to decide whether a JIT system is feasible. Runners and regular repeaters can be a useful area of application.

Restrictions on storage quantity are often based on value. The high-value goods are restricted to the bare minimum required to enable the production line to function. At the other end of the spectrum, the low-value parts can be stocked in larger quantities and topped up at longer time intervals. The effect of these time periods is that some movements occur regularly in the JIT system but others only once or twice per day. It is not necessary to keep everything consistently on the move unless economics dictate, so the method is selective.

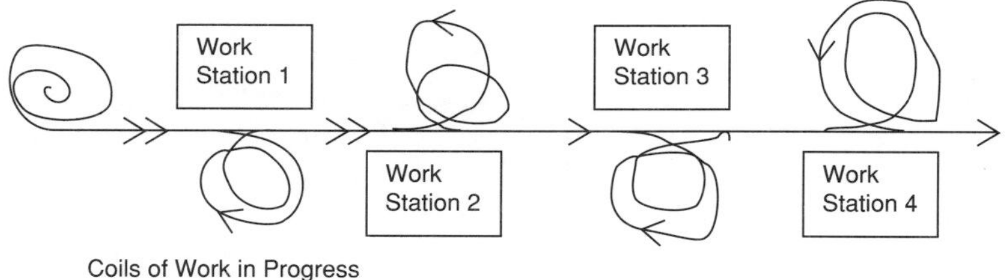

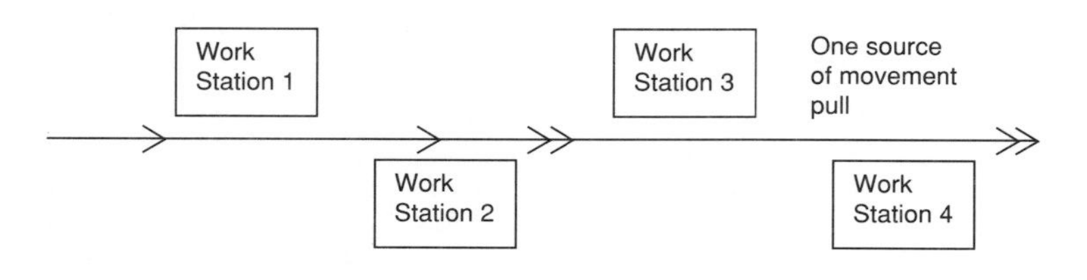

Figure 7.3 Effect of push and pull systems

Basics of Method Study

Method study is an investigative procedure to assist in the analysis of identified problems. The six steps of method study are (see also Chapter 6):

1. Select work to be studied. Reasons why this particular job has been chosen are enumerated.
2. Record the present method. Using diagrams or charts the present system is recorded on the basis of five symbols plus numerical data:
   O operation which adds value, such as Cut, Mix, Assemble, Pack, Process
   ⇒ movement or transportation ***adds** cost*
   □ inspection ***adds** cost*
   ▽ official storage ***adds** cost*
   D delay occurs ***adds** cost*
   Activities that add cost: Moving, Counting, Finding, Chasing, Storing, Reworking, Kitting, Inspecting, Recalling, Recording.
3. Examine the present method. Subject each stage to the (1H, 5W) test, that it How?, Why?, When?, What?, Where?, Who? questions.
4. Develop a better method. Use analysis, experience, consultation and techniques to make savings and gain agreement.
5. Install the new method. Nothing is achieved until this stage is successfully carried out. Plan, prepare, train, obtain materials and tools, and then explain.

6. Maintain the new method. Savings are immediate but can also be lost immediately. Monitor the process to detect any changes made for whatever reason, which by definition should be harmful.

The assembly line needs this kind of method study analysis to assist in work balancing, but also in the organisation and planning of the changeover from one product to another.

One of the golden rules of method study analysis is to produce, whenever possible, simultaneous activity, whether by individuals or groups, in preference to sequential activity. This is obviously a great time-saver. For example, two activities taking 9 minutes each can be performed in either 9 minutes simultaneously or 18 minutes sequentially. This idea has ramifications for the changeover method presently used.

See references in Chapter 6 for more details of method study techniques.

D. Effects of Applying 'Just In Time' to an Assembly Line

Introduction

The assembly of fluorescent light fittings, using a work-balanced team of operators and a power-conveyor system for transportation, would not appear at first sight to be a fertile ground for large savings. However, at Heartfords Ltd there was a problem of not consistently achieving output to the standard rate that the line was designed for. It was planned that 60 light fittings would be assembled and tested every hour. This target was achieved for an hour or two but various interruptions brought the weekly output figures well below expectations. Many ideas had been put forward and various monitoring trials of all activities were made. Unfortunately, each investigation highlighted different reasons for the delays, for example, shortages of various components, but some were made in-house while some were bought from outside suppliers. Also staff problems were identified, such as absenteeism, regular self-certification or time spent training new operators. Changeover problems persisted for each different product assembled, including lost tools, poor-quality parts, testing unreliability, packing problems and so the list grew and grew. The present method had been designed and planned about four years previously; it was time for a rethink and a new style of investigation that would produce action. Looking for a 'quick fix' had not been successful, hence the basic methods needed to be questioned.

Joanne Walters' success in redirecting the extrusion department's efforts towards a high degree of performance encouraged the new General Manager to invest in her time once more. Although the problems were in a different part of the factory, Joanne decided to use the same accepted formula which had been so successful in unravelling the complexities of extrusion. Another graduate engineer, Tom Scholes, who had about 8 months' experience of Heartfords' operations was included in a small team along with the stores and assembly department managers. Tom was to work full-time at different positions on the assembly line for 4 weeks, and then monitor the areas agreed with Joanne and the managers. Similar tactics to those used previously were employed and new data from the investigation were obtained.

Present Position Outlined

A range of different light fittings is assembled on each of eight powered, specialised conveyor-lines. The time necessary for each production run is typically about one day to one and a half days, which translates into about thirty changeovers of product per week spread over the eight lines in the department. The quantities produced range from 200 to 1500, so flexible production scheduling is required to keep the assembly lines active and the customers satisfied. The scheduling is partly helped by assigning about ten products, of which three or four are regularly demanded, to each line to enable familiarity of product assembly specification to be established with operators and supervision.

The conveyor-lines are set out back-to-back to save space (see Figure 7.4). However, much of the space available is required for component parts for assembly which are placed near each operator station. The light fittings are bulky but not heavy, and so are easy to handle by the women operators.

Extracts from Tom's reports: mode of assembly

The first step towards customer order satisfaction occurs in the finished part stores. About two days before a particular product is scheduled to be assembled on a particular line, action has to start in the stores. A complete kit list of all the purchased parts necessary to fulfil the scheduled order, say 600 light fittings, is provided by using the computer terminal files and a standard stock list of items. Personnel in the stores identify parts, collect them from nominated areas, count them into boxes or plastic bags, seal them and put them, suitably labelled, into a large pallet trolley. If, say, 1200 screws are required, then a box of 5000 may be opened and the 1200 removed by the stores worker for assembly on the previously mentioned 600 order. If weighing is used for the count, then the box of screws would be taken to a power point from which weighing scales can operate. If not, counting, estimating or guesswork may be used. This was suspected, if not proven.

When this activity is completed the excess screws are returned to the nominated storage place. Generally, one person is assigned to collect a full kit, however, when customer demands dictate, the worker may leave one order partly completed to fulfil a more urgent one, sometimes with another worker, because of a changed or slipped schedule. When the assemblies are

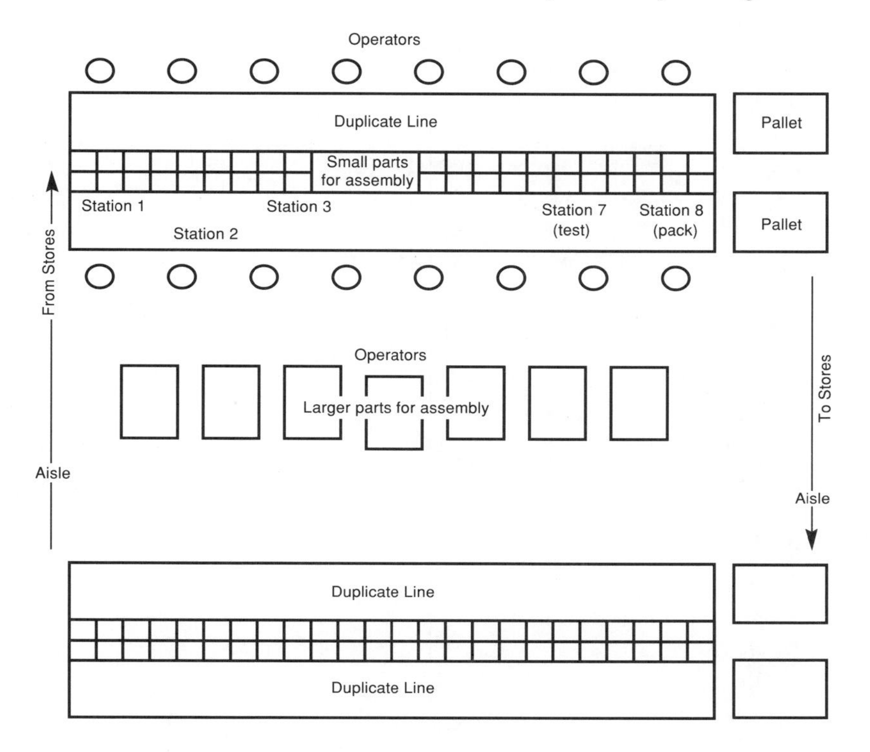

Figure 7.4 Plan view of double assembly line; one of five such layouts

required, sometimes the accurately completed kit of parts is pushed on its pallet trolley to the nominated assembly line. The delivery may either be to the expected schedule or sometime later if delays have dominated the store's activities. These delays originate from the management not adhering to the monthly schedule, as agreed, but choosing to respond to market changes which are more volatile than the monthly increments currently used in planning can accommodate.

When the previous job is completed on the line, which is taken to be when the last assembly has cleared all operations, the outcome is a line with all operators waiting for components, tools and instructions. To make the line ready for the next product assembly requires the supervisor to unpack the kit of components provided. Each part unpacked is allocated to a particular station based on the supervisors' prior knowledge and experience. Some mistakes are made with the allocations at this time. Meanwhile the operators are removing excess parts (kitting errors) left from the previous job, changing tools and fixtures to suit the new job. If any of the tools are missing or incomplete, a search of other lines is carried out to make good the shortfall. In theory, all the excess parts should be collected by the line supervisor and returned to the stores to enable store workers to return them to the appropriate storage area. This does not happen.

Eventually, each station is refurbished with components and tools. Only then can station 1 build the first part of the assembly while all the other stations wait. The work from station 1, when completed, is transferred to station 2 and two operators are now working. This process continues progressively until finally all stations are working. Any problems identified during these early stages are generally corrected by the line supervisor without large delays to others. A considerable time can elapse before a good product is produced, as a result of the changeover. More difficult problems mean extra-supervisory activity off the line, and all stations have to stand still while this occurs. Visits to the stores by the supervisor or making contact with appropriate offices for information again cause considerable delays. When the assembly of 600 products is complete, the whole process of changeover with its delays is repeated for the next job. This activity occurs about 1500 times per year; any unnecessary delays experienced are expensive and wasteful of capacity, that is potential output.

Problems Identified by 'Bottom Up' Analysis

As a last resource, the management had decided to create a special task force to investigate and solve the continuous difficulties with the eight assembly lines. This was now critical because the rise in demand for certain light fittings was beyond the output achievement of the company.

The consultant from the Group Service Organisation, Joanne Walters, was to lead a small team of three, chosen from the areas of stores, industrial engineering (Tom Scholes) and production assembly. The team was given four weeks to find out the real problems and four weeks to develop solutions and report the costs and timescale required.

Joanne Walters was particularly keen not to use the company's standard data collection system, but to find a source of primary data that could be relied upon and have it backed up with different reasons for each delay highlighted.

The new approach adopted to investigate the operation of the lines involved one member of the team working as an operator on one of the lines (Tom Scholes) and recording reasons for each delay observed. Problems can be analysed by either a 'top down', method, examining why management decisions are not carried out as they filter down the system, or 'bottom up'. The latter involves identifying all inefficiencies at shopfloor level and determining how management above must change to remove these problems that it helped to create. In this case 'bottom up' methodology was used, mainly because 'top down' had failed in the past from lack of commitment to the limited findings.

The main areas of concern identified were:

1. The changeover between production runs was taking too long, often over one hour, which meant the equivalent of one production line out of the eight was completely lost to effective output.
2. Shortages in the complete kit of parts occurred, resulting in stopped production lines from 5 to 12 per cent of the running time. Delays occurred because more parts had to be obtained from the stores or, even worse, fresh orders had to be placed for parts that were not available at all. These delays also almost equated to one equivalent line of production being lost. Lines were kept waiting after completion of one order because the kit for the next product was still in the stores and incomplete; time was lost while harassed stores workers, when available, were assigned to complete it, producing further errors.
3. Tools were missing as a result of a policy of sharing tools, such as nut runners, screw drivers and soldering irons, across the eight lines in the department. This inevitably led to some of the operators hoarding popular tools to assist their individual line performance against the rest of the assembly department requirements.
4. Excess parts were numerous at the end of each run but the 'booking back procedure' to the stores was rarely used. With the pressures of organising changeover and start up, plus the need for fast output of assemblies, the booking back system is seen as an unhelpful activity to supervisors.
5. Workers were also using one of two unofficial 'solutions':
   (a) throw away all excess parts as scrap to clear the area (what are a few screws worth?), or
   (b) keep parts in hidden storage around the line in case they were needed later as a shortage item. This could create problems in space usage and incorrect part identification, resulting in the wrong small parts fixed to an assembly.

A survey of parts thrown away or lost to the system showed that the value was between £20 and £200 for each changeover. From the samples taken the average value lost to the company was £100.

As approximately 1500 changeovers per year can take place, the final figure of loss was estimated to be in the order of £150,000. Most of this money could be saved if an alternative system could be found to retain these parts. This would also have dramatic effects on the accuracy of inventory levels.

E. Questions for Reader and References

Consider the advantages and applications of Just In Time (Kanban) with reduced set-up time methods and the problems of accurate kitting of assemblies. What should be done? Give reasons for your ideas.

Possibilities of action

Several alternatives have already been proven to be unsuitable. A whole range of solutions derived from the 'top down' management approach has had little effect. Industrial Engineering Department personnel have tried in vain to modify each part of the system on a bit-by-bit approach. These failures have necessitated the new investigation by the task force using the 'bottom up' approach, which should lead, via the data collected, to the development of solutions for the delays identified.

The possibilities are therefore built around acceptable changes in material identification and flow, plus a new way to reduce the changeover difficulties. Often finding the solution to one set of problems means that another set of problems that are of concern are easier to tackle. The questions have been framed around possible solutions; it is the 'how to achieve it in detail' alternatives that are now under consideration.

Questions

1. Can a Kanban philosophy help in this case?
2. Can space be saved by using Kanban methods?
3. Does the changeover of the line have to be sequential?
4. Can changeovers be achieved in a simultaneous mode?
5. Will accurate kitting or Kanban save the company up to £150,000 from discarded surplus parts?
6. Can changes be made that assist the stores operators to do their jobs more effectively?
7. Can schedule adherence be improved by changes you wish to make?

Suggested references for further reading

Advanced Manufacturing Technology Management, M. Harrison, Pitman Publishing (1990).

Manufacturing Systems, edited by V. Bignell *et al.*, Blackwell (1985).

Operations Management – Concepts, Methods and Strategies, M. Vonderembse and G. White, 2nd edn, West, USA (1990).

Production and Operations Management, A.P. Muhlemann, J.S. Oakland and K. Lockyer, 6th edn, Pitman Publishing (1992).

F. Proposals Adopted After Task Force Recommendations

Kitting was considered to be the key to many of the problems identified, so the qualities of Kanban were tested. The recommendation was to substitute a new, regular and small quantity refurbishment of the components system for the present individual counting and kitting of specific orders. The parts

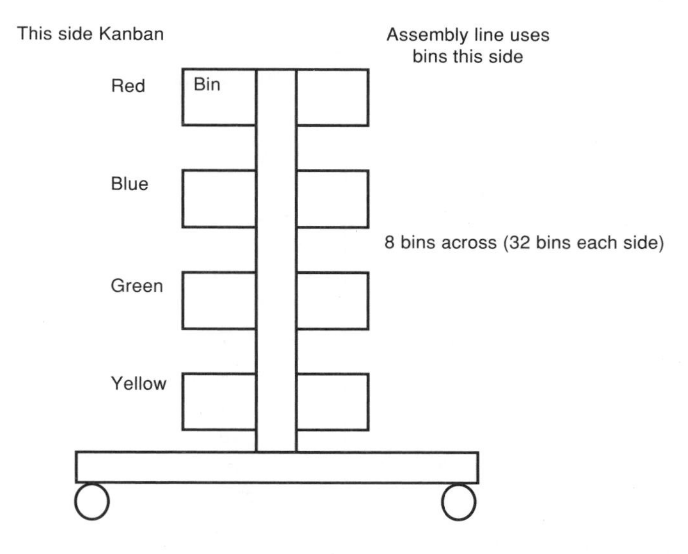

Figure 7.5 Specially adapted trolley for ensuring a continuous supply of bought out parts. Quantities checked twice per day

lists used for assembly of all light fittings on one particular assembly line were analysed for feasibility with Kanban (runners and repeaters). Of the typical 36 components per assembly, 8 were major parts made in-house at the company, and the rest were 'bought out' finished parts supplied from outside sources. Internal company difficulties prevented, at this time, the Kanban system being applied to these major 8 parts, but the remaining 38 typical parts could be subjected to the Kanban system of parts control because of cross-usage of the same parts. Standardisation does help. To enable this to work, special storage trolleys incorporating a removable two-bin system were adapted from existing equipment, and containers were purchased (see Figure 7.5). The adapted trolley was designed to be loaded with 'bought out' finished parts before assembly changeover was due to commence. The trolley was to be wheeled into place, when required, as a provider of assembly storage bins and to act as a central storage point. The second of the two storage bins provided would then be put on to the assembly production line for easy access by the assemblers. The first set of bins would be retained on one side of the special trolley with the bins displaying the part number in a clearly visible style. When the assembly line bin was empty, it would be exchanged with the full one on the central located trolley. Twice per shift the trolley would be checked, so all empty bins could be recharged.

This procedure uses standard quantities of component parts that obey the following two criteria:

1. A standard-order quantity or pack size is supplied (no counting).
2. The usage rate is such that one loaded bin will last for over half a shift.

If this will not work, the bin is recharged with more than one standard quantity. Recharging empty bins is not a time-consuming job so this frees stores workers to kit the 8 major parts made in-house. At the end of each assembly schedule, when all required assemblies have been produced, the bins on both the assembly line and special trolley contain excess parts as a result of the Kanban. All bins are put on to the trolley and wheeled to a special section of the stores. Later the individual bins are withdrawn, when required, and put on to another trolley of parts required for a particular assembly line output.

This situation of reduced variety was greatly helped by identifying multiple applications across the product range of the typical 28 components (screws, washers etc.), and these could then be stored in an appropriate multiple-application zone of the stores. The Kanban list would also alert the design office to the advantage of using parts from this multiple-application zone for the design of future light fittings.

The throwing away of 'unofficial' storage and left-over components was eliminated by the use of the Kanban system. Purchasing of component quantities from outside suppliers was rationalised by ordering standard amounts that could be subdivided or multiplied into the recharge quantity that each bin required for each Kanban round.

The outcome meant that orders were placed for 200, 500, 1000, 2000 and 5000, which were the standard recharge quantities for the empty bin. When a large light-fitting order was received, the above quantities could be ordered from the suppliers in multiples of the figures given above.

Next the jig and tools supply system was changed. Tools were matched to the product range produced on each line and stored underneath it, so the tools were always available when required. Each tool was marked clearly to identify its particular place of usage and storage. All jigs and fixtures needed were treated in a similar way. This did mean duplicating some tools and eliminating or refurbishing some others. These costs were easily justified as potential savings would come from increased capacity/output as the many delays were reduced. Finding the tools needed when the old system was used was further complicated because of the obstructive nature of some individuals protecting their own collection of company tools; this activity was now obsolete.

Finally, the changeover method was altered completely. The new system was greatly assisted by all the previous changes, particularly the Kanban of component parts and tooling availability. As the last sub-assembly was completed at station 1 and was moved to station 2, the first operator started to clear the unwanted Kanban bins on to one side of the central trolley. Then station 1 could be reloaded with parts from the newly arrived adapted storage trolley for the next products. Localised tools were changed on the overhead air lines and the fixtures put on to the conveyor bench. The light-fittings major components were already kitted in pallets for the whole assembly line and put alongside the appropriate stations. Using this method only about 10 minutes elapsed before the first stage of the new assembly was completed at station 1 and journeyed down the line. This procedure continued until complete tested light fittings were put on to a stillage ready for despatch.

The simultaneous changeover method now introduced meant that delay for each worker was only in the order of 10 minutes compared with over 70 minutes taken by the sequential system. The extra capacity gained by this saving was equivalent to having an extra assembly line for no extra cost! Comprehensive individual station instruction sheets for each new product were supplied and greatly assisted the smooth transition from one product to another. Operators did not have to rely on their memory and experience alone. Some operators had limited experience anyway.

The results achieved from these changes increased outputs by up to 10 per cent above the original output, which was not normally consistently achieved. The real comparison was the monthly output figures, these were 40 per cent above the previous system's best performance and consistently obtained. Predictions of output achievement, from the production control department, were now more accurate, which then made the schedule adherence close to 100 per cent. Production scheduling was changed to a weekly plan instead of the unreliable changeable monthly plan. This became feasible because the overall system was now more responsive without the large delays and inconsistencies of the past. Alterations of schedule were not necessary any longer, as everything could be scheduled into the plan for next week at the latest.

The increase in output could be attributed to the almost total elimination of waiting for tools to replenish missing components and time lost during the slow changeover of products. A standard performance indicator system that the company used to measure outputs demonstrated a step change for the better immediately the new methods were applied to each assembly line in turn. Indicators, such as operator performance, ineffective or lost time, output per hour and quality levels, all improved to the highest figures for any department in the company.

Summary

A lesson to be learned from this application of new ideas to the department is the interrelationship of each identified problem. Improvements in changeover times between products would have only been marginal if delays were still occurring in kitting, part shortages and tooling availability.

The whole system has to be seen as an integrated activity operating within the company strategy. Spin-offs naturally occur for the whole system when a line becomes more effective. For example, in the stores area the reduction of parts to be counted into kits for the eight assembly lines meant that other identified problems in the stores could have the extra resources available applied directly to them. Also the two daily tours to restock the Kanbans were always made on time because the regular store panics and resulting unplanned actions have been almost eliminated. Note also the simplicity of each of the solutions; the best way is always the easiest one, and does not add complications to an already chaotic situation. Cut out the waste and delay to keep the essential features of activity visual rather than relying on memory or recorded data to control a process. This new assembly system is highly visual, and control can be exercised by a supervisor observing situations as they happen in each area. The kind of essential visual checks that can be made include the following:

1. Are the Kanban bins full?
2. Are the Kanbans being replenished at the right times?
3. Is the changeover between products being carried out to plan?
4. Are the documented instructions being used?
5. Is output per hour consistent and on target? Use an automatic counting display board to make results always visible.
6. Are tools available and operational?
7. Is the area cluttered with excess parts?
8. Have regular discussion groups been set up to encourage operator feedback?
9. Have the workers been encouraged to report the problems concerning the operational control at an early stage, so that prevention can be quick and effective? Fast response encourages further feedback. The opposite is also true.

8 New Investment Decisions

Section

A. Learning Outcomes

No equipment lasts forever, at least in economic terms, so the management of a company must decide to make provision for new and replacement operations. As technology progresses, the capital sums and risks involved increase so the methods of the past are not adequate for the situation of today. To tackle decision making in a systematic way requires factual information and not opinion and feelings. Business acumen is still in existence but it needs the risks analysed and cash flows estimated to make it work.

This study will demonstrate that:

1. Even good decisions regarding equipment purchase are ineffective if the operational control is missing.
2. More than one person/department often need to get involved with the investment and, without good liaison, errors in specification can occur.
3. Investments are often based primarily on cost considerations so all the problems discussed previously concerning costs are present.
4. Good decision making includes installation and maintenance of the correct operating procedures; these include reporting and monitoring to set achievable targets.
5. The problems of investment analysis are critically investigated to show the pitfalls that exist, primarily in the area of cost and cash return modelling.

B. Learning Objectives

After studying and analysing this case study presentation the reader should be able to:

1. Explain the steps in investment analysis.
2. Identify the key activities in decision making.
3. Define different types of investment criteria.
4. Differentiate between the case for investment and the ability to make the investment give the required return.
5. Identify critical areas of data collection both before and after the investment is installed.
6. Define the essentials of fast 'start up' after installation to minimise unnecessary costs.

C. The Basics of Investment Analysis

An overview of different decision criteria is undertaken to contrast and compare their relative effectiveness in decision making. A number of ways are available to enable investment decisions to be made, each having its strengths and weaknesses, but all rely on accurate data.

1. Payback Period

This is a popular and simple way to justify a decision. It is based on estimating the surplus for each year's activity of the investment and then summing up the yearly increments to establish when the total equals the value of the investment, using interpolation for yearly proportions of part years.

Example

		Project A	*Sum total*	*Project B*	*Sum total*
Investment	£	(100,000)		(60,000)	
Year 1		20,000	20,000	10,000	10,000
2	I	40,000	60,000	15,000	25,000
3	N	40,000	100,000	20,000	45,000
	C				
	O		3 years		3¾ years
4	M	40,000		20,000	65,000
5	E	50,000		30,000	

(*Note*: Figures in parentheses are negative.)

In both of these cases a comparison can be made and a conclusion drawn that Project A 'pays back' before Project B (3 years to 3¾ years): certainly quick and simple. However, the result depends on the accuracy of forecasting the savings, the method ignores the value of payments beyond the payback point and finally it treats all money values as the same; that is, £40,000 in 3 years from now is equivalent to £40,000 received now. The method is popular but its shortcomings need to be recognised. Many companies set a time that they will not exceed for the payback period. A popular target is 3 years, however, depending on the cash flow condition of the company, it can vary from as low as 9 months to 6 years in many instances.

2. Return On Investment (ROI)

This method does use all the future returns on investment that the company considers will be achieved. The benefits could go on for as long as 20 years in some cases, however, this is most unlikely to be considered. It is rare for projects to be analysed over more than 5 or 6 years. Anything received after this period is a bonus and in some instances it then becomes what is called a 'cash cow', something that can be 'milked' for as long as possible.

Take the previous example.

	Project A	*Project B*
Investment £	(100,000)	(60,000)
Year 1	20,000	10,000
I 2	40,000	15,000
N 3	40,000	20,000
C 4	40,000	20,000
O 5	50,000	30,000
M		
E Profit	£90,000	£35,000
Average profit	$\frac{£90,000}{5}$	$\frac{£35,000}{5}$
	= **£18,000**	= **£7,000**

$$\text{Return On Investment (ROI)} = \frac{\text{Average profit}}{\text{Capital invested}} \times 100$$

$$\text{Project A} = \frac{£18{,}000}{100{,}000} \times 100 \qquad \text{Project B} = \frac{£7{,}000 \times 100}{60{,}000}$$

$$= \mathbf{18\ per\ cent} \qquad = \mathbf{11.7\ per\ cent}$$

Conclude A is better than B!

Sometimes Average Return On Investment is used instead of ROI to accommodate the notion of the investment being depreciated down to zero over the life of the project, that is start at £100,000 and end at zero in 5 years. Hence, average return should not be based upon £100,000 but the mean of £50,000. The result of these changes is a doubling of the ROI. For example

$$\textit{Project A} \quad \frac{£18{,}000}{100{,}000/2} \times 100 \qquad \textit{Project B} \quad \frac{£7{,}000}{60{,}000/2} \times 100$$

$$= \mathbf{36\ per\ cent} \qquad = \mathbf{23.4\ per\ cent}$$

Flattering but not helpful to clarify decision risks.

When figures are quoted for ROI it is sensible to ask how they have been calculated. The answer received can be highly instructive. The problems are similar to the payback period method, but this does have the advantage of considering all the receipts, not just those up to the payback point.

If the payments for the example quoted had been in the reverse order the criteria become:

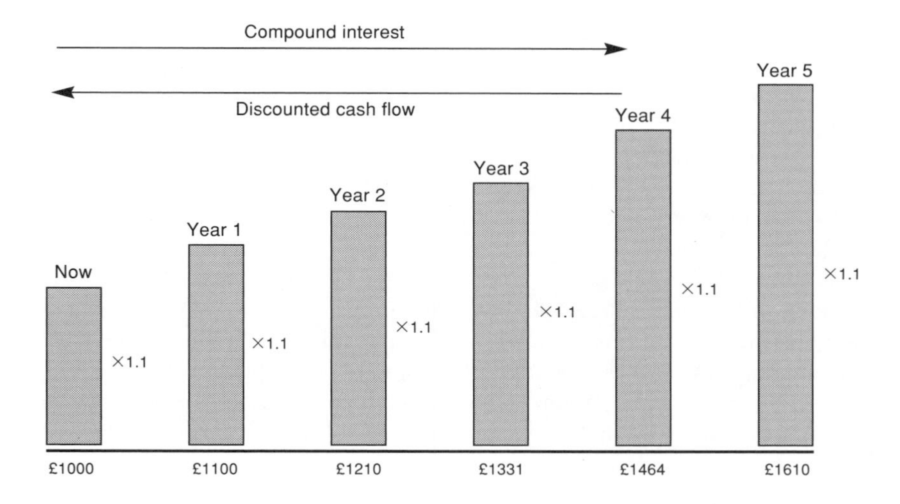

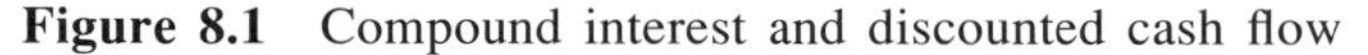
Figure 8.1 Compound interest and discounted cash flow

		Project A	*Sum total*	*Project B*	*Sum total*
Investment £		(100,000)		(60,000)	
I	Year 1	50,000	50,000	30,000	30,000
N	2	40,000	40,000	20,000	50,000
C			$2\frac{1}{4}$ years		$2\frac{1}{2}$ years
O	3	40,000	130,000	20,000	70,000
M	4	40,000		15,000	
E	5	20,000		10,000	
Profit		90,000		35,000	

Payback is different but ROI is still the same.

This leads decision makers to consider both Payback and ROI together, which can lead to a conflict of criteria.

3. Discounted Cash Flow

This method is designed to take into account the time value of money, so the problem of whether £20,000 today is equivalent to £20,000 in 2 years' time is not incorporated. The time value of money is understood by people but the theoretical background to the calculation must be established.

Compound interest is directly proportional to Discounted Cash Flow (DCF). Take £1000 and invest it at 10 per cent in a Building Society. The investment would grow as illustrated in Figure 8.1.

Hence if the time value of money is 10 per cent the relative value of the £1000 over each time period is as shown. (Interest rates can be different from the time value of money.) £1000 now is the same as £1100 next year or £1210 in 2 years time etc.

The time value of money relates to the penalty that companies incur if the cash flow is delayed. Different companies choose to use a figure that suits their present position and this helps to judge the risks in investments.

The concept of DCF is the opposite to interest; for example, £1464 in four years from now is equivalent to £1000 today or £1100 next year. By this simple mechanism the total can be found by adding together the amount for each year based on the value of money today; this is called the Net Present Value (NPV).

Compound interest is calculated by using a multiplier of $(1 + i)^r$ where i = interest factor and discount factor r = number of years. However, the DCF requires the opposite, that is use $(1 + i)^r$ as a *divider* to establish the overall PV value. For example £1464 in 4 years' time has a PV of

$$\frac{£1464}{(1 + i)^r} = \frac{£1464}{(1 + 0.1)^4} = \mathbf{£1000}$$

It is normal for the factor $\frac{1}{(1 + i)^r}$ to be given in tables as a fraction, for instance

$$\frac{1}{(1 + 0.1)^4} = 0.683$$

and the calculation of NPV becomes

$$£1464 \times 0.683 = \mathbf{£1000}$$

Applying the NPV criterion to the previous example, the value of money again being 10 per cent:

		DCF	*Project A*	*PV*	*Project B*	
Investment		1	£(1,000,000)	(100,000)	£(60,000)	(60,000)
Year	1	0.909	20,000	18,180	10,000	9,090
I	2	0.828	40,000	33,040	15,000	12,390
N	3	0.75	40,000	30,000	20,000	15,000
C	4	0.683	40,000	27,320	20,000	13,660
O	5	0.62	50,000	31,000	30,000	18,600
M						
E	**Total NPV**			**39,540**	**£8740**	
	NPV × 100 = **40 per cent**					= **14.6 per cent**
Investment over 5 years						

A larger difference is then demonstrated to reinforce the demands of Project A.

Yield for DCF

The present value (PV) method is useful but limited, a better predictor from the viewpoint of rationality is the yield. In this method a set time value of money is not chosen or assumed but it is found by search (trial and error). The yield or calculated time value of money is the figure at which the profit (that is, return in excess of investment) growth is cancelled out by DCF. This sometimes needs a little extrapolation to achieve and is best demonstrated by the previous example. *Note*: 10 per cent leaves a residue of £39,000 so start at, say, 25 per cent and search.

	Project A	*Try 25 per cent*		*Try 20 per cent*	
Investment	£(100,000)	1	(100,000)	1	(1,000,000)
Year 1	20,000	0.8	16,000	0.833	16,660
I 2	40,000	0.64	25,600	0.694	27,760
N 3	40,000	0.512	20,480	0.578	23,120
C 4	40,000	0.41	16,400	0.482	19,280
O 5	50,000	0.328	16,400	0.4	20,000
M					
E			**94,880**		**106,820**
			(5,120)		6,820

25 per cent too high, try 20 per cent and this is too low.

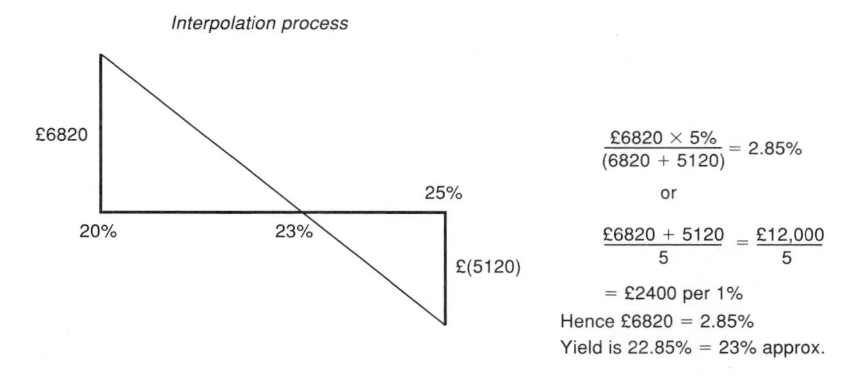

Figure 8.2 The use of interpolation to gain accuracy from reasonable estimates

In the same way the yield for Project B is calculated. 10 per cent gave an excess of £8740 so try **15 per cent**.

		Project B	*Try 15 per cent*	
Investment		£(60,000)	1	(60,000)
I	Year 1	10,000	0.87	8,700
N	2	15,000	0.756	11,340
C	3	20,000	0.657	13,140
O	4	20,000	0.571	11,420
M	5	30,000	0.497	15,000
E				59,600
				(400)

This is near enough, hence, yield is **15 per cent**. Yield is sometimes referred to as IRR (Internal Rate of Return).

Although the returns in Projects A + B have been discounted, it is really the same as an increase when viewed into the future. In this case Project A has a yield or growth of 23 per cent because it took that degree of discounting to reduce the surplus to zero. Another way of seeing the completeness of the yield figure is to demonstrate the yield as a provider of cash flows as the capital reduces to zero. The typical effect of investment, can be visualised as a 'black box machine' with a lid through which we can put money in and take it out. Also a revolving handle that multiplies the money in the box at a set rate each year. If the box is taken as Project A then £100,000 is put in and the box should then generate five years' cash flow as it multiplies by the previously calculated 22.85 per cent yield.

Search is used to enable a reasonable estimate to be found. Interpolation can then be used to give greater accuracy (see Figure 8.2).

'Black box' simulation

		£	£
Year 0	**In**	(100,000)	**Out**
Year 1 becomes		122,850	20,000
Cash left		102,850	
Year 2 becomes		126,350	40,000
Cash left		86,350	
Year 3 becomes		106,080	40,000
Cash left		66,080	
Year 4 becomes		81,180	40,000
Cash left		41,180	
Year 4 becomes		50,000	**50,000**
Cash left		Zero	

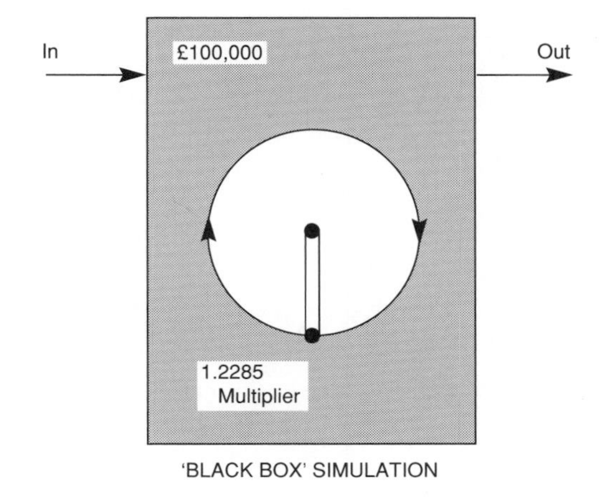

Figure 8.3 Generation of cash flow from investment and multiplier (yield)

This is a model of Project A generating the expected cash flow by having a suitable multiplier built into the system (see Figure 8.3). The knowledge of the size of the multiplier assists the decision making because it correctly mirrors what is expected to happen. None of the other method's figures can duplicate this model cash flow and so their investment rates are suspect and misleading. A final word of warning: each system relies on the validity of the cash flow estimated from sales and activity costs. There is no special method available to eliminate the errors built in by poor estimates. The old saying, Garbage In Garbage Out, applies. To establish each of the yearly positive cash flows generated by a project's activity means using the detailed costing models discussed in Chapter 5. Essentially, for simplicity, each year can be considered the same, that is next year will be repeated. So sales figures and selling prices produce revenue while the costing figures can be assigned by absorption of overheads to different factors or marginal contributions which include depreciation of the new asset to be purchased. If the investment is for a narrow range of units then the figures for all of these can be included, however, if the work is of a more general nature, then typical figures of activities or other generalisations for total outputs are used.

Much detail is involved to obtain accurate figures, and files of data and analysis are produced, but as the proposal moves from level to level towards the Board of Directors it often becomes more and more condensed. It is not unknown for the final distillation of data to consist of one page of A4 with grand totals under a limited range of headings to establish a 'Yes' decision by whatever criteria the company uses. This is the style Perfecta Engineering used and an outline pro-forma is shown in Figure 8.4.

Value added per year by investment	300,000
Costs Fixed costs applied to area to be used , X m^2, at £Y per m^2	100,000
Installation cost – resale value of plant divided by Z years	50,000
Power requirements per year	15,000
Labour costs per year (including superannuation etc.)	15,000
Factory OH applied to area to be used, X m^2 at £R per m^2	50,00
Surplus/profit	70,000

$$\frac{\text{Profit} \times 100}{\text{Installation} + \text{capital costs}} = \frac{70{,}000 \times 100}{200{,}000} = \textbf{35 per cent return}$$

Recommendation: Yes, No, Postpone

SIGNED:

Figure 8.4 Capital Expenditure Form (CEF)

D. Present Position at Perfecta Engineering

Perfecta Engineering produced an important part in their marketing range which, in different forms, became a major component in all assemblies. The main operation in the production of this part, a transducer, was the accurate machining of the structure from a solid piece of aluminium. To produce this important item a special cell had been designed which incorporated four standard conventional machine tools around a vertical CNC Machining Centre. The conventional machines were used in the preparation of the aluminium block using the operations of milling, drilling and grinding. The operations helped to facilitate the CNC Machining Centre activities, reducing the time required on the machine significantly. A further bonus of this cell design is that these conventional machines could be operated while the automatic cycle of the CNC machine was working, hence only one person was required to operate the cell which, theoretically, could produce at a rate of about four completely machined transducer structures per hour. This cellular system had now been running for over 18 months of which about 12 months had been on a full two-shift work pattern. The first 6 months had been used for familiarisation and training before the expected outputs were scheduled (an example of a slow start-up). However, in practice all had not gone well, the output figures on a weekly basis were inconsistent and lower than expectations by a wide margin. Supervision had been asked to deal with the matter and to correct any problems detected. The feedback information received by management from the various supervisors was disappointing and unhelpful. It involved a series of excuses, such as problems with jigs and fixtures, swarf removal, breakdowns, absenteeism, quality of material used and poor tooling. The answers varied according to whom management spoke and which day they asked the questions. All this confusion resulted in no action being taken and eventually a steady weekly output of about 100 to 120 units per shift was achieved instead of the expected feasible target of 160 units per shift. The CNC Machining Centre was programmed to achieve this figure (160) given 100 per cent utilisation of the operating time. Extra pressure put on to this disappointing performance was an increasing order book with shorter lead time demand from customers. The two shifts could not meet the needs of assembly for the transducers, so the company took the decision to subcontract all the extra demand outside. Eventually, considerably more transducer units were made externally than those made internally per month.

Any pressure to change this state of affairs was muted because, according to figures provided by the accounts department, the cost to make was about £5 more than the cost to purchase. A new works manager was appointed at this time and when he eventually analysed some of the facts of the situation he decided to contact the Head Office of the Group and ask for assistance from the Industrial Engineers Consultancy Service.

The consultant appointed to review the problem area was Hugo DeWitt who was familiar with operational control difficulties. His brief consisted essentially of two questions:

1. Why is the internal cost £5 more than the subcontractor's price? (Subcontractor is profitable at this cost, internally no profit made at the higher price.)
2. Can the external work be moved back to Perfecta Engineering with either a duplicated cell or some other production system to increase the company's contributions towards its profit margin? (Something that buying from outside could not achieve.)

Initially he had the problem of little data availability, except output figures and scrap rates plus a few uninformed opinions. He had to decide how to proceed before he could answer the questions set in the necessary factual manner. The lack of information meant that Hugo would have to arrange for data collection himself. The firm had no shopfloor data collection system whatsoever, so the culture to assist the collection did not exist. Preliminary discussions with supervision and operators were not encouraging and an initial manual collection system, needing the co-operation and commitment of the workforce, did not work well and was stopped.

Start-Up Losses Reduce Productivity

New equipment can be a straight replacement for a similar operation or can be a new technology in a new area of operation. Whatever the degree of difficulty, replacement should be understood, planned and controlled especially during the critical period of the first few months of operation. This is a make or break time for investment success, a time when operating standards, attitudes and achievements are formed. Failure to grasp the importance of this period by all management can cause intractable problems in the future. Hugo DeWitt had experienced many times the difficult specification of needing to sort out practices that have been developed for more than 12 months. He had often been called in after failure had developed. Optimism of better things to come is the main reason for management inaction. When action was eventually considered then inevitably outside help is sought through consultants like Hugo, showing the distance management likes to put between themselves and the situation it was responsible for creating. History had showed that simply connecting new technology to the required services, training the operator at the supplier's works then providing one or two typical jobs for assessment under controlled conditions was not a recipe for complete success. Handing over the responsibility to ill-informed, that is untrained, supervision for the immediate start-up arrangements of the project would not ensure success. It is useful to remind oneself that the paperwork for the investment decision is based on a full year of achieving the targets set. If a poor start is made and progress to the set targets is slow, then much time will elapse before the cash flow generated is up to planned expectations. Hugo could give various examples of this situation, ranging from 2 years to never before the cash flow was acceptable.

Successful start-up needs to be planned and monitored, and corrective action taken immediately to demonstrate management's interest and commitment for the new venture. Anything less will be interpreted as either 'it does not matter anyway', or 'the management is in agreement with our performance', or 'for all the shouting and anger the management throws at us it actually does nothing to change the situation'. None of these attitudes or similar ones should be allowed to develop, and the only way to prevent it is for Hugo and others to be involved until success is clearly seen by all concerned. Once the successful pattern of work has been established the excuse of not having ever achieved targets that the management has set becomes invalid. To explain more clearly, Hugo DeWitt produced Figure 8.5 to demonstrate what had happened during the first cell's start-up and what his monitoring had highlighted.

Section 'A' on the graph gives the strategic organisational loss in that the company does not expect to use all the 168 hours available in a week. Output and maintenance needs dominate these strategic decisions.

Section 'B' shows the losses planned from start-up, as the utilisation of

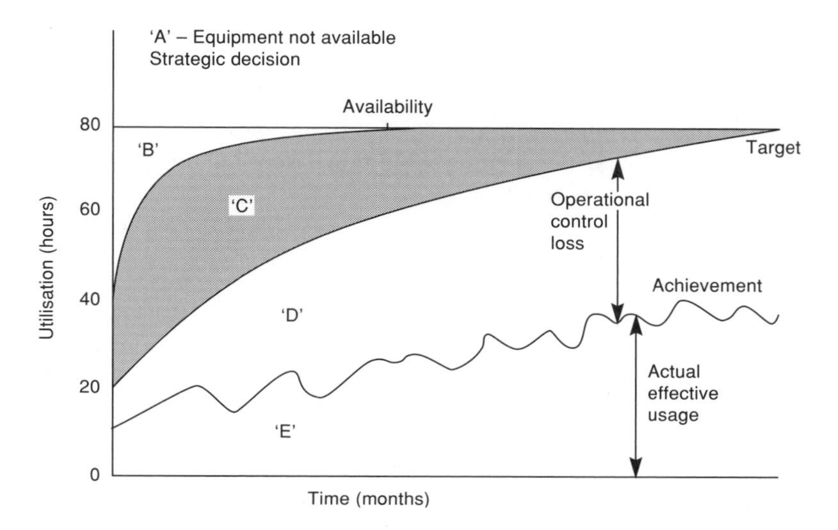

Figure 8.5 Start-up effectiveness and long-term losses

the cell moves from part single shift through overtime to two-shift operation.

Section 'C' allows for a moving upward expectation with time, because learning, confidence and repeatability will eliminate the transitional losses. Then, in a planned timescale, installation can be considered to be complete and operational to company standards.

Section 'D' shows the losses from set targets experienced because of all the reasons previously quoted, essentially well under 75 per cent achievement of targets.

Section 'E' highlights the monitoring information obtained and the resulting lack of success. The losses are ongoing and cumulative, they are not just a particular week's failure but a continuous failure, even if a fluctuating one. The area of 'operational control loss' represents months of lost output. The graph demonstrates this clearly and can be a useful tool in change management discussions. Reasons for these losses, as mentioned before, are many and varied, but inaction by responsible management is at the centre of the situation. As Hugo pointed out, he had to find what was happening because the management could not tell him with any certainty. An example of this was the CNC machine, which proved unreliable during the first 6 months, and supplier service engineers regularly came to make the equipment operational again under the warranty. This satisfied all concerned until a particular order was considerably delayed. A Director of Perfecta Engineering. after making a quick investigation wanted compensation from the manufacturers for the lost production that the delays caused by non-availability of the CNC machine. No record of these delays existed, only the dates they had sent for the engineers. He then set up a monitoring system, which did not reveal further long delays but did reveal a pattern of failure, and when the identified root cause was eliminated the machine reliability was excellent. Information is the base on which sound decisions are made, otherwise guess work is involved, as just demonstrated. However, these errors will no doubt be repeated as, once the machine settled down, the monitoring process was abandoned.

First Actions Taken

Hugo decided to use an automatic system of collecting data directly from the machines in the cell. To achieve this goal, sensors were placed on the various pieces of equipment in the cell to record any change of activity by sending a signal to a computer. Examples of these sensors were:

1. movement of the swarf removal conveyor;
2. movement of the spindle of the drilling machine;
3. movement of the spindle of the milling machine;
4. movement of the table of the grinding machine;
5. movement of the guards on the CNC machine;
6. whether the autocycle light was 'on' or 'off' on the CNC Machining Centre.

All these were signal inputs to a program which identified the changes against a date and timescale. Also output summaries of each day or shift performance could be obtained and diagrammatic representations of results printed. The computer was stored in a locked box near the cell, and each floppy disc used could store a full week's activity, which meant little supervision of the system was required. After 6 weeks of analysis, Hugo found accurately the proportions of time for each of the real reasons for delay. Previous cases he had investigated showed that surprises could be expected. These sets of results were no exception and included:

1. Breakdowns of the CNC machine were very low but the swarf conveyor system was unreliable and had caused losses in production of between 10 and 15 per cent. A close examination highlighted its unsuitability for removing fine aluminium swarf because the interlocking leaves of the steel belt became clogged, as did the drive mechanism. A replacement belt was necessary in these circumstances, which would need to be continuous and not provide an opportunity for swarf impregnation.
2. Although the running time of the CNC program was 13 minutes and easily identifiable in the data collected, between the 13-minute cycles a highly variable time span occurred. The unload/load activity was not supposed to take long but sometimes there was a considerable delay between one cycle finishing and the next one starting. It was also established that conventional machines in the cell were being operated while

the CNC machine was standing still on about 40 per cent of the delay occasions; the rest of the delay occurred when no activity at all was recorded in the cell. A further insight into these facts was that the night-shift operator produced more parts than the day-shift operator. The acquired knowledge of the local supervision had made the team think the opposite. It was convinced that the nightworker was slow and not very effective. The only reason he had not been replaced was because no one else had been trained to operate the CNC machine. More problems and delay arose on the day shift than on the night shift, but absenteeism was higher on the night shift.

3. Further investigation revealed that part of the delay on changeover from one finished component to the next one starting was due to the complicated cleaning process required to the fixture before a new block of aluminium could be loaded. The fundamental problem was the 'vertical' design of the machine, which meant that all the fine swarf produced stayed in the area around the work. The coolant flow was insufficient to wash it away. Awkward recesses in the fixture also helped to trap and retain the swarf. Modifications were then carried out that helped swarf removal, however the initial planning sheets had not provided time for a cleaning activity at the unload/load stage. This activity had become an accepted part of the job although it took time.

Another surprising feature brought to light was the spread of small delays from 20 seconds or so to 5 minutes. They were liberally sprinkled throughout the control cycle of the CNC machine, especially on the day shift. Hugo realised that the operator, for some reason, would not let the machine run under the control of its program for long periods. The sum total of these small delays became a large percentage of total lost production. They mainly go unnoticed in normal visual monitoring because the expectation of the ending of the delay is always present. It is operator and management optimism that rules out the detection of these apparently insignificant delays.

The results also revealed that the CNC machine was far from being the management-controlled operation situation that was expected after paying £60,000 for the Control System. The operator was seen to be the main cause for delay contributing to the resultant longer cycle time. Computer control of machine movements via a controlled fixed program means, in theory, that cycle consistency is readily available. Performance should be different from processes where the manual input is the dominant controller. Data taken by Hugo DeWitt gave a spread of results that confirmed his analysis, namely the operator was dominant in both situations. Hence, the spread of performance for manual control and that of CNC were highly similar. This meant that the advantage of the Control System was being lost. Figure 8.6 outlines the extent of the problem and includes frequency expectation to highlight the degree of mismatch.

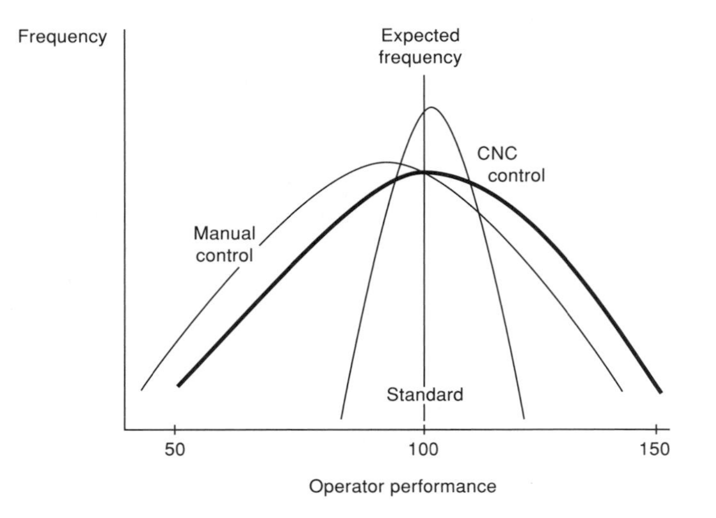

Figure 8.6 Results of control level obtained when CNC control is installed

The CNC Control System should have been centred, with little spread, on the 100 standard. Performance figures of 140/150 should not be obtained if the times allowed were completely realistic. Questions put to the supervisor of the department could not be answered with any conviction because the supervisor knew little about the CNC Machine's operation, since he had had no training in that area. This contrasted with the rest of the equipment in his responsibility, on which he was something of an expert. A gap in management knowledge is not uncommon when high technological equipment is purchased.

When Hugo started writing his report to the works manager, he decided to highlight his findings and point out the areas of potential savings over and above the modifications to the system that he had already reported on.

Investment decisions are notoriously variable in their degree of success and as someone once said: 'It is always a bad time to invest, excuses can always be found to do nothing'. However, what is the cost to the company of doing nothing? These and many more questions needed answering before the final decision could be made. He realised that the management of Perfecta Engineering had a strategic decision to make. They had three options, which would be laid out in the report clearly; however, he would not find it hard to choose between these options. The difficulty for him was whether the management would see the situation as clearly as he did himself; some politics were involved, as usual, and this causes people to think in ways that cannot be easily justified to all.

The three options outlined consisted of:

1. Make small changes to the existing cell to improve performance above the best so far achieved (that is, 120 units per shift) and keep ordering the major part of demand from the subcontractor as at present.
2. Change and duplicate the existing cells to use them as competitors/comparisons to break through some of the fixed cultural attitudes to output achievement.
3. Start from the basics of demand to consider which is the best way to produce the units with the new knowledge and experience gained by monitoring performance. This would mean taking the outside price and quantity as the norm, and justifying a different way of producing the units that would stand as a realistic investment on its own.

The Third Option Chosen

After a month of debate, a decision was reached to form a team of three people to investigate the feasibility of Option 3 and to make the best case they could for not purchasing subcontractor company units. The team consisted of Hugo, plus an accountant and an undergraduate engineer on a sandwich placement. They split the work between themselves after agreeing a step-by-step plan to formulate the report in 5 weeks maximum.

Initial Team Findings

In addition to the previous information given, certain other facts concerning potential new cells were obtained. A horizontal CNC machine suitable for the work involved was available at a cost of £300,000 which can have fitted to it an automatic pallet changer for £50,000. This would enable the load/unload cycle to be performed while the machine was working. The activity cycle of the machine could, theoretically, then be almost 100 per cent. Cycle times verified by the manufacturer for manual load/unload and machining cycle were 12.0 minutes. For the pallet changing automated cycle it was 8.5 minutes. The advantage of a horizontal machine is that the swarf falls away from the cutters and workpiece, unlike the vertical machines used at present. Only one person would be required to operate the new cell; all the other jobs in the cell could be done well within the 8.5 or 12 minute cycles. If extra fixtures were purchased then two components could be loaded at a time, which would double the cycle time and allow less activity by the operator. The extra fixtures cost £4000 each. The new cell would require 40 m^2 of space, which could be found alongside the present cell. However, on the other side of the department there was a suitable area available of 45 m^2. These areas were vacant or had the occasional pallet stored for a period. None of the team expected a 100 per cent performance and considered that the investment decision should be based on a conservative, less-risk strategy based on the potential output. Sales would present no difficulties in the new situation as 30 000 units were supplied by the subcontractor each year and were purchased for £40 each. All the figures previously used can be assumed to apply for the new feasibility study.

Finally, further calculations requested from the finance office revealed that the present cell used 30 m^2, and that the recovery rate for fixed costs was £305 per m^2 per month and for factory overheads was £140 per m^2 per month.
Note: By successfully using a greater area of the workshop, some people thought that these figures would be reduced. Others thought that the area was 'free' because its costs were already accounted for. The decision is yours!

E. Questions for Reader and References

1. What is the current position with regard to planned output? Is a surplus or loss now generated?
2. Use marginal costing to verify cost per piece for both 'make' and 'buy' strategies.
3. Investigate the Horizontal Machine in both modes, that is Semi-Automatic or Fully Automatic, to test for feasibility. Is the new space required (not used) important financially in this feasibility analysis?
4. Is the fully automatic cycle necessary because the operator is not fully utilised? Give reasons for your answer.
5. (a) Choose an investment analysis method, after verifying each one previously detailed. Which is appropriate in this situation? Use 75 per cent as a performance target per shift rather than the theoretical 100 per cent.
   (b) What is the advantage and disadvantage of using 75 per cent?
   (c) If 90 per cent was achieved in practice, what difference to profits per year would this make?

Note: The company figures for overhead recovery already quoted can be used, if required, in future investment analysis.

Suggested references for further reading

Accounting in a Business Context, A. Berry and R. Jarvis, Chapman & Hall (1991).
Advanced Manufacturing Technology Management, M. Harrison, Pitman Publishing (1990).
Introduction to Cost and Management Accounting, R. Storey, Macmillan (1995).
Justifying Investment in AMT, IEE/CIMA, Kogan Page (1992).
Management Accounting for Decision Makers, G. Mott, Pitman Publishing (1991).

F. Proposal Adopted by Perfecta Engineering

Team and Company Actions

The team's submission to management was based on the following recommendations:

1. Invest in a new cell, taking note of previous errors, and bring in the subcontracted orders to provide the extra workload. It is important that the costs of the units from the new cell are competitive with the subcontractors price of £40, not the internal customer price of £47.50.
2. The horizontal machine with the auto pallet loader was also found to be the better choice. Although the operator was not overloaded, this is not a reason to reject the auto pallet loader because the extra output obtained will justify the extra investment.
3. The team were unhappy about charging the 40 m^2 of space to the investment when this space was not carrying overheads at present. All the costs normally allocated by the conventional accounting system would provide additional contribution to the company and these should not be allowed to have a negative effect on the decision of whether to invest or not. However, to please some directors etc. in Perfecta Engineering the report quoted costings based on cost per m^2 because the cash flow was still acceptable.
4. The report started with an analysis of the present position regarding costs. Allocations used by the accounts department transferred well into the report.

Company Report on Transducer Units (old method)
Using Absorption/Historical Costing

Rate of production 3 per hour
Cost of labour £6 per hour
OH on direct labour 600%, that is £36
Cost of labour and OH = £42 per hour inclusive

Cost per unit $= \frac{£42}{3} =$ £14 each for material, £21 per unit
but scrap rate increased cost to £23

Depreciation of CNC cell
£81,000 per year

Output per year 240 per week × 45 = 10 800

$$\text{Depreciation per unit} = \frac{£81{,}000}{10\,800} = £7.50$$

Cost to manufacture

	£
Labour + OH	14.00
Material	23.00
Depreciation	7.50
	44.50

Internal 'selling price' £47.50 **£3 profit**

Cost to buy £40 each from subcontractor
Profit £7.50

Completed by team to demonstrate errors

CEF for proposed first cell investment
Value added 320/week × 45 weeks = 14 400 output
£(47.5 − 23) = 24.50 (14 400 × 24.5) = £3,528,000

Costs	£
Fixed costs 30 m^2 × £305 per month per m^2 × 12 months	110,000
Depr. **£240,000**/3 years	80,000
Power & exps (Maintenance Agreement)	20,000
Labour operators (2 × £12,500)	25,000
Factory overheads 30 m^2 × £140 per month per m^2 × 12 months	50,000
Total cost	**285,000**
Profit	**67,000**

$$\frac{\text{Profit}}{\text{Capital}} \times 100 = \text{ROI} = \frac{67.8}{240} = \textbf{28 per cent}$$

ROI acceptable.

CEF for Actual First Cell Investment Based on Data Obtained

Output at 270/week (best result) for 45 weeks (optimistic)

	£
New value added 270 × 45 × £24.50	298,000
Some savings on power (£2000) due to inactivity (as before)	
Total cost	283,000
Profit	**15,000**

Based on internal selling-price of £47.50 each

$$\text{ROI} = \frac{\text{Profit}}{\text{Capital}} \times 100 = \frac{£15{,}000}{£240{,}000} \times 100 = 6 \text{ per cent}$$

Does purchase from outside for £40 give a better return? *Note*: DCF not used by Perfecta Engineering. Also note: Some companies have such a poor costing system and a demand for high rates of return over a short period of payback that apparently 'uneconomic' investments are refused. One way of checking the validity of costing investments by internal accountants is to obtain information on the costs of leasing the equipment rather than purchase. The hourly costs are then easily obtainable and can be compared. This can make 'anti-leasers' think again on the bases of their own costing figures concerning the 'cheaper' alternative of ownership.

CEF 75 per cent Utilisation for New Automatic (preferred choice)

Firm's standard method used

	£
Value added 18 900 units at £(40 − 21) = £19 × 18 900	359,100
Fixed costs 40 m^2 at £305 × 12 months	146,400
Depreciation $\frac{£360{,}000}{3}$	120,000
Labour (2 operators)	25,000
Factory overheads 40 m^2 × 140 × 12 months	67,200
	£378,600

Less (£19,500)
Do Not Proceed

Some other case needs to be made if this 'good' investment is not to fail.

Marginal Costing Approach Adopted

Output per shift/yr	5400
Wages per shift/yr	£12,500
Power & expenses	£10,000

$$\text{Marginal cost per unit} = \frac{£22{,}500}{5400} = £4.20 \text{ (+ material cost £23)}$$

$$= \mathbf{£27.20}$$

Internal sales £47.50 each hence
Contribution is 47.50 − 27.20 = **£20.30**

$$\text{Breakeven output} = \frac{\text{Fixed costs}}{\text{Contribution unit}}$$

$$= \frac{\text{Depreciation} + £160{,}000\ \text{OH}}{£20.30}$$

$$= \frac{81{,}000 + £160{,}000}{£20.30} = \frac{£241{,}000}{£20.3}$$

$$= 11\,871 \text{ units needed}$$

Two shifts provide for 10 800 (max.), so contributions cannot cover allocated OH.

Test Calculations (100 per cent utilisation)

£40 target for total costs

Semi Automatic 5 units/hour
Fully Automatic 7 units/hour

		Depreciation
Capital costs	Semi Automatic £ 300,000	100 000
	Fully Automatic £ 360,000	120 000

Fixed costs (as before)

£110,000	space allocated overheads
£ 50,000	space allocated overheads
£160,000 +	£20,000 + £25,000
	Power Wages

Output per week 5 × 80 = **400** Semi Automatic
7 × 80 = **560** Fully Automatic

Output per year at 100 per cent Semi Automatic 18 000 [Test at 75 per cent utilisation]
Fully Automatic 25 200

Less scrap produced Material cost £21

£40 − £21 = £19 value added.

Total marginal costs = £22,500 as before

Marginal costs per unit

Semi Automatic $\frac{£22{,}500}{18\,000\ \text{Output}} = £1.25$

Contribution = £19 − £1.25
= £17.75 per unit

Fully Automatic $\frac{£22,500}{25,200 \text{ Output}} = £0.9$

Contribution $= £19 - £0.9$
$= \underline{£18.1}$

Depreciation + Fixed costs
£100,000 + £160,000 = **£260,000**
(not changed)

Breakeven quantity, Semi Automatic (£ 17.75) = $\underline{£260,000}$ = 14 650

Profit = (18 000 − 14 650) £17.75 = 3350 × £17.75
= £**$\underline{60\,000}$**

$$\text{ROI} = \frac{\text{Return}}{\text{Capital}} \times 100 = \frac{£60,000}{£300,000} \times 100 = \underline{20 \text{ per cent}}$$

Breakeven Quantity (Auto)

Overheads = £160,000 (space) + £120,000 (depreciation)
= $\underline{£280,000}$

$$\text{Breakeven} = \frac{£280,000}{£18.1} = 15\,450 \text{ units}$$

Profit from (25 200 − 15 450) £18.1 = £**$\underline{176,500}$**

$$\text{ROI} = \frac{\text{Return}}{\text{Capital}} \times 100 = \frac{£176,500}{£360,000} \times 100 = \underline{49 \text{ per cent}}$$

Recommendation
Purchase Auto Pallet Changer and bring subcontracted orders back to company.

Test Calculations for 75 per cent Utilisation and 90 per cent

Semi Auto 18 000 units × 75 per cent =
13 500 per year/(90 per cent) = 16 200 (repeat)
Auto 25 200 units × 75 per cent =
18 900 per year/(90 per cent) = 22 680 (repeat)

Cash Flow Semi Auto $\frac{£22,500}{13\,500} = £1.70$ marginal cost

Contribution = £19 − £1.70 = $\underline{£17.30}$

Auto $\frac{£22,500}{18\,900} = £1.20$ marginal cost

Contribution = £19 − £1.20 = $\underline{£17.80}$

Semi Auto cash flow = £17.3 × 13 500 = £230,000
Automatic cash flow = £17.8 × 18 900 = £330,000

Allow for £160,000 overheads (Management Demand)

Semi Auto	*Auto*
£230,000 − £160,000 = £70,000	£330,000 − £160,000 = £170,000

	Semi Auto		*Auto*	
	75%	*90%*	*75%*	*90%*
Investment	£300,000		£360,000	
Year 1	70,000	120,000	170,000	248,000
2	70,000	120,000	170,000	248,000
3	70,000	120,000	170,000	248,000
4	70,000	120,000	170,000	248,000
5	70,000	120,000	170,000	248,000
	£50,000	£300,000	£490,000	£880,000
Average	£10,000	£60,000	£98,000	£176,000

$$\text{ROI} = \frac{£10{,}000}{£300{,}000} \times 100 = \mathbf{3\%/18\%} \qquad \frac{£98{,}000}{£360{,}000} \times 100 = \mathbf{27\%/50\%}$$

Floor area has become more effectively used. Great benefits from achieving 90 per cent utilisation, especially on the Automatic Machine.

DCF PV and Yield for 75 per cent Utilisation

Not considered by management but provided by Hugo's team.

	Semi Auto £	Try 6%		*Auto* £	Try 40%	Try 35%
Investment	(300,000)	1.0	(300,000)	(360,000)		
Year 1	70,000	0.943	66,000	170,000		
2	70,000	0.89	62,300	170,000	Results required	
3	70,000	0.84	58,800	170,000		
4	70,000	0.792	55,400	170,000		
5	70,000	0.747	52,300	170,000		
	Over-discounted		(4,300)		(14,050)	17,400

Say 5 per cent yield

Use extrapolation $\frac{14{,}050 + 17{,}400}{5}$ = £6290 per percentage point

For Automatic

$$\frac{17{,}400}{6290} = 3 \text{ percentage points extra}$$

Answer 35 + 3 = **38 per cent yield**

Note: Perfecta Engineering did not use DCF because the Financial Director claims that decisions must be clear cut, that is obviously 'yes' or the answer is 'no'. He thought that clarifying the sometimes 'grey areas' was not helpful: an idea that Hugo's team could not accept.

These figures confirmed the team's views.

Discussion

The marginal costing approach was a difficult argument for Hugo's team to win, especially as it meant overheads for the new area were to be ignored. This was against everything that the present costing system was based upon.

Finally, the decision was taken at Board of Director level to go ahead with the proposal, that is Horizontal Machine with pallet changes, because it was seen as viable even with the overheads included. Also taken on board was the need to provide a 'cell champion' who would be responsible for ensuring success. It was soon decided that the space on the other side of the department would be used to ensure the negative attitudes of the present cell would not be too close and formative. To further assist this idea, extra operators were recruited so that they could be trained in the new methods without the conflict of tradition. Use of internal personnel was considered to be too closely identified with resistance to change. The trade union representatives supported expansion of the workforce, seeing it as a vote of confidence by the company in bringing back the subcontracted work. A paper-based monitoring system was designed by the 'cell champion' to record outputs per hour, delays over 5 minutes with reasons, to isolate scrap or to rework for analysis. Feedback information would be conveyed to management on any reason for output falling below target. Computer recording was discontinued at the request of management. Bonus payments were introduced to provide incentives for achievement above the targets set, even though the cell was CNC program driven, because it was recognised that its success depended on good communications between operators' machines and management. The close involvement of the 'cell champion' meant that each difficulty was quickly addressed. While 100 per cent supervision was not likely, the very facts that close interest was shown and when necessary fast action was taken consistently and reliably, encouraged the operator to cell-champion communication.

From the initial start-up, the economic minimum target level of 75 per cent utilisation was achieved after *two* weeks, and one month later the 88 per cent level was attained and generally maintained. The resultant cash flows generated meant, according to the company's new marginal cost approach, that the payback period was now 18 months and resultant profits were the highest generated anywhere in the sub-group. One of the enigmas of all this change and success was that very little improvement was obtained in the original cell even though some of the new operational style was attempted. The 'traditions' and expectations of operators, supervision and the technology used could not transform the situation that they were in. In fact, the most experienced operator in this cell left to take up a similar job, nearer his home, for a £100 per week rise. One of the problems encountered in this group company was poor pay and the resulting attitudes of the people who chose to work there or found they could not leave for obvious reasons.

Summary

A lesson to be seen from this investment analysis and installation is in the insecure ground that most of the decisions and actions were founded upon. A clear analysis of the present situation and the advantages of change needs to be made. The investment decisions were based on essentially the changes made on a cost versus the extra revenues generated. If all other costs are ignored (overheads), then the analysis is concerned with gaining an overall improvement in the situation at minimum risk, that is contribution maximisation.

Three questions that can and should be asked about any new investment once established are:

1. Does the new investment generate more sales revenue?
2. Have the stock levels (including work in progress) reduced as a result of the investment?
3. Have the yearly costs of operation reduced as a result of the investment?

Most investments can give an answer of 'yes' to one of the questions, but what about the other two? The aim should be to gain from all three areas; this is the challenge to investment decisions and the *post-mortem* of installations. Companies often claim success in a generalised way. However, when confronted with these specific questions and having the need to substantiate their replies, the claims of success can quickly melt away. In the case of the new cell, all the three questions were answered in the affirmative and figures were available to prove these responses.

Finally, some companies do not even conduct a *post-mortem* in which success or otherwise is evaluated, because they prefer optimism to experience. Some do conduct a one-day *post-mortem* analysis, about 6 months after start-up, with advanced notice and the knowledge given to all that it will not be repeated. Surely success can be guaranteed for one day! The benefits of the second cell came from continuous monitoring and correction of errors, making a *post-mortem* unnecessary.

9 Division of Labour in a Radiator Repair Business

Section

A. Learning Outcomes

Should organisations and methods be allowed to develop naturally or is a scientific input necessary to shake people out of their optimism and complacency?

This case study is based on throughput quantity and hence delivery time to meet customer requirements. The complacency demonstrated where a metal product is concerned would not be tolerated in an industry where shelf life was short. Why should company managers think so differently when the bottom line required result is the same.

This case study will demonstrate that:

1. Releasing work into a manufacturing environment with requested delivery times does not guarantee success.
2. When management refuses to see the problems because it cannot face change, then outside help is needed.
3. The success of the past does not create success now if the conditions have changed dramatically.
4. Lack of knowledge concerning activities on the shopfloor, operation time and delays, means production is not under control.

B. Learning Objectives

After studying and analysing this case study the reader should be able to:

1. Explain the impact of the division of labour when capacity increases are necessary.
2. Identify key features of delay reduction and control for management.
3. Define capacity and scheduling needs to allow target setting and achievement to be readily available.
4. Understand the problems of working in an environment based on continuous failure to achieve customers' overall needs.
5. Differentiate between changes that are fast and low cost to install, with those that require time and investment.

C. The Basics of Work and Work Flow

Introduction

An old very experienced engineer once said '*There are only two problems in industry: work movement and people.*'

This simplistic statement should not be immediately rejected because it demonstrates some truth, particularly in the case outlined.

Work movement problems are integral in this case study together with the impact of changes on people. The major situation comes from management's attitude and inertia.

Layouts

Material can be processed or manufactured in a number of ways.

Method 1

In a fixed position, that is, materials, tools and manpower are brought to one place. Examples include ship building, bridges and power stations. Some sub-assembly can take place elsewhere.

Method 2

Use of the 'High Street' shopping model means moving products from process to process, each of which is located in set position, and each one a specialised process: symbolically, greengrocer, butcher, chemist etc. Components move through a series of workshops following a planned route to suit their needs. The route can be long, chaotic and wasteful because of the queues that form. Examples include small batch production of tools (Chapter 4), specialised electric motors, dies and moulds. This is normally called '*job shop*' or '*process*' layout.

Method 3

Use of the 'Supermarket' shopping model means moving material to a set position of process resources, all in one area, which gives closer control, more local responsibility and fewer queues. The engineering term for such a localised layout is *cellular manufacture*.

Method 4

Use of a 'Flowline' layout means a predetermined, no-option route incorporating all that one component needs and always required in large volumes. Examples include car engine blocks, refrigerator assembly, car body assembly and universal joint manufacture.

Sometimes a method exists in the basic state but hybrids are everywhere as engineers try to design the best system for each manufacturing need. Many factors are included in the search for best method.

The previous ideas built around 'runners, repeaters and strangers' (Chapter 4) still apply in work flow and layout. In this case study the problems are centred around 'strangers' and occasional 'repeaters'. An extra dimension involved is the batch size, which is generally only **one** and rarely more than **three**.

When a problematic situation exists in the workflow through operations, then data and evidence need to be either available or, if not, collected. The data must not consist of what people think is happening but established factual data of what is happening at this time. The usual method of study techniques could be applied, as in some previous chapters, but this area of enquiry is very different.

In this case the emphasis is on the importance of logical questioning to reveal the facts and allow analysis. Observing and recording all that happens between 'material in' to 'material out' is still necessary and needs to be put on a diagram or chart so that problems are shown clearly. However, if a dispute is known to exist within management concerning the facts, further in-depth study is often required to answer all the questions posed by the conflict.

Method study provides a comprehensive questioning analysis in one of its six steps. Essentially the investigators find the answers over time to set questions, not however by asking the listed questions to anyone but by finding the answers themselves. The set questions are useful for obtaining information about Person, Purpose, Place, Position and How each operation is done. This procedure gives a condensed outline of potential change available in any situation. A whole range of supplementary questions needs to be asked and answered to enable the investigator to answer the set questions completely. The sum of all this effort can be incorporated on to a preprinted form called a Critical Examination Form (Figure 9.1) for analysis and record purposes.

To obtain useful data an interview agenda, completely informal and known by the experienced investigation practitioner, is used to tackle the problem at different levels of detail. Normally these move from the general to the particular. This method is revealing because the level of accurate knowledge known by the interviewee is discovered at a specific level. From this information, or lack of it, the investigator must take appropriate action for the data collection. Asking detailed questions straightaway is popular as a one-step approach, but the theme and thrust can be lost when a series of general answers are received. By structuring the questions from the general to the particular the interviewer can stay in control. The opposite is true when other techniques are used.

The Interview's Informal Agenda

If information is needed, say, in the productivity improvement area of a company, then ONE way of attempting this broad area is by obtaining answers to these typical questions.

General question

'Which is the main constraint on increased productivity – Planning or Control? Different questions are needed to test the chosen reply (Planning, Control, Do Not Know). For example, if Planning is quoted, a number of routes can be investigated as follows.

Answer: Planning Route 1 suggestions
'Are delivery dates competitive and achievable?' If 'yes', ask what evidence is there? If 'no', investigate further; for example, ask 'Where do most delays occur?' and 'What have you done or are doing to improve the situation?'

A RECORD OF ANALYSIS FOR:

JOB: INVESTIGATOR: PROCESS: DATE:

Factors	*Basic questions*	*Answers*	*Possible Alternatives*	*Recommended alternative*
Purpose	What is achieved? Is it necessary?			
Place	Where is it done? Where else could it be done?			
People	Who does it? Can anyone else do it?			
Sequence	When is it done? Can sequence be changed?			
How	What is the method used? Can it be changed?			

Figure 9.1 Critical examination form

Answer: Planning Route 2 suggestions
'Are all process cycle times known and reliable?' If 'yes', move to another route. If 'no' ask: 'Is work measurement used to set target times?' also 'Is your equipment comparable with your main competitors?'

Answer: Planning Route 3 suggestions
'Are your forecasting and scheduling reliable?' If 'yes', request evidence and move to another route. If 'no', ask more questions:
'Who makes the forecasting and scheduling decisions?' 'What is the percentage of rush jobs in a typical month?' 'Are you happy about this position?'

A similar style of questioning is used if the Control option is chosen to try and find in detail the number of issues that constitute Control failure. In this specific case study, the work flow delays were identified as key elements in output failure.

Diagrams and charts recording work movements through the necessary processes are extremely useful for systematic analysis (see Figure 9.4 later in the chapter). They are intended to reflect the typical activity in the workplace; however, as variations in operations are common, it is not easy to establish what is typical. A period of time that does not include untypical occasions is necessary, such as carrying out checks on absenteeism when transport strikes are happening. Charts do not show absenteeism anyway, so other such data need to be collected elsewhere. The charts are used for recording specific activities that happen to a chosen product or material as it is processed, and consist of:

1. Operation – a change of state occurs.
2. Transport – material is moved.
3. Inspection – checks are carried out.
4. Storage – material is in a planned store.
5. Delay – material or manpower is held up by unplanned activities.

The aim of good layout and methods to produce effective work flow is the elimination of delays and waiting of all kinds. The customer normally pays for the operations, so more operations completed mean more sales receipts. This cannot be said of storage, inspection, delay and transport activities. On some occasions, operations might consist of only 20 per cent or less of all activity. Some firms target 40 per cent operational activity to measure effective work flow.

Delay elimination is considered in the alternatives available. In Figure 9.1, one of these changes is based upon the 'division of labour'. The investigator must ask whether an alteration in work allocation will be beneficial.

Labour is not either 'divided or undivided' under this heading, it is a continuum that stretches from one person doing every job activity possible, for instance Sales–Manufacture–Delivery to Customer. The other end of the spectrum is job size cut down to a few seconds and many hundreds of people involved doing a short specialised task.

A one-person business is a good example of one side of the range and will nearly always finish in some short task like putting two wheels on one side of a car on an assembly track. General investigators establish where a particular job is on this scale and consider whether a change in either direction is helpful. One direction gives *job enlargement*, the other way *divides* the job further and makes it less skilled with shorter training and more repetitive actions over the shift. In the recent past the short cycle repetitive work was considered desirable and the most theoretically efficient. However, practice in this area has not always produced the output figures expected with and increase in operating costs. A range of markers on the continuum of the division of labour showing the theoretical advantages is given in Figure 9.2.

Moving from left to right in Figure 9.2 shows the influence of the 'division of labour'; each section displays the increase in productivity expected per person. Product demand of considerable proportions is needed to justify each step shown. The cost of each unit drops the nearer the right-hand side is approached. Note that the theoretical basis of expected outputs is not always achieved because counter-influences operate in work areas that demand short cycle repetitive work.

An example that demonstrates this dilemma is an assembly line output. This example is based on real output statistics. The assembly line was designed to make 50 units per hour with 200 workers, but they rarely met the target. Assembly often stopped because of mechanical/electrical failure in one part of the long line or through personnel disputes, quality problems and temporary absenteeism. The situation was analysed and second thoughts brought a different design configuration.

The end result was four shorter assembly lines, each with 50 workers, with a target output of 10 units per hour. While it was now planned to make 40 units as against 50 units, certain advantages arose from the job enlargement exercise:

1. Targets were consistently achieved and sometimes bettered. (Theoretically 50, rarely achieved, means poor delivery promises, but a certain 40 plus gives reliable output to satisfy customers.)
2. When breakdowns and disputes happened they were localised to one line, thereby marginalising the effects; for example, 30 units per hour were still being produced.
3. Communication with a team of 50 workers is easier than with 200. Agreement is naturally a factor of group size.
4. Assembly cycle time increase means fewer repetitive actions but more skills are required. This provides a more satisfying work situation.

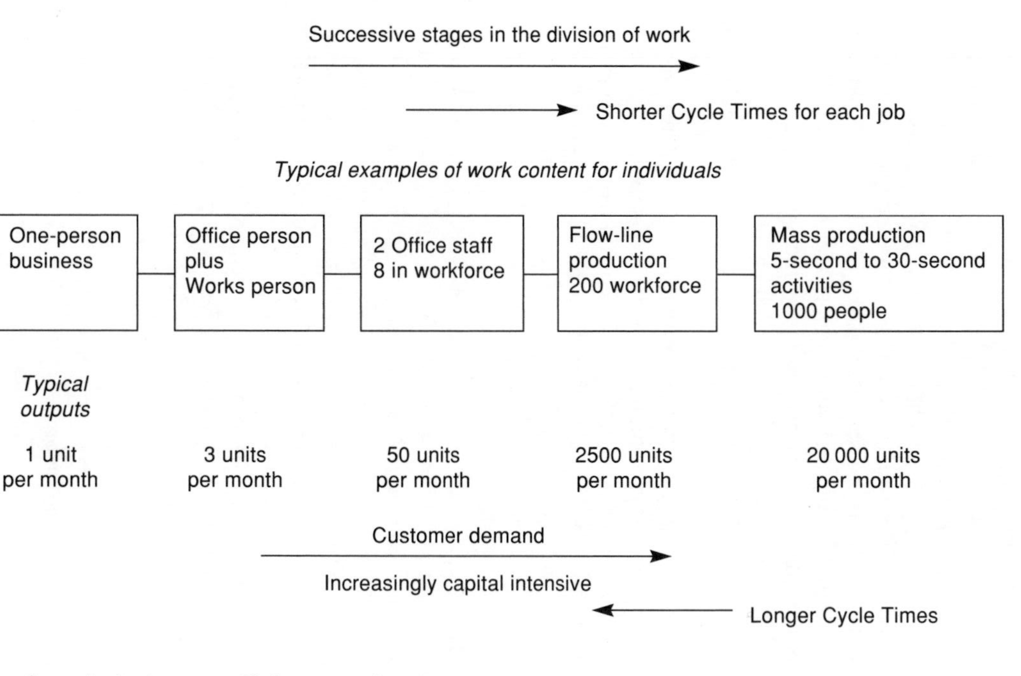

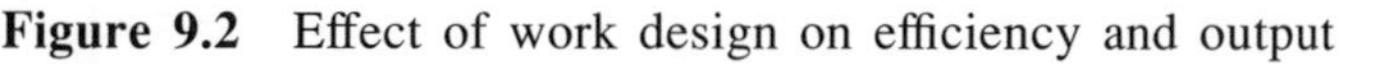

Figure 9.2 Effect of work design on efficiency and output

5. Statistics showed that the theoretical 50 units per hour output often dropped to 20 units or fewer when morale was low, for whatever reason.

These operational decisions on layout and work practices give a pointer to how and why it is necessary for industrial engineers to match the worker and company expectations. Good work flow is just one part of the many-faceted manufacturing service industry designed inputs. Without it delay and cost increase are inevitable. Many factors can combine to cancel out a potentially good layout for workflow. Assembly lines cannot be operated successfully if absenteeism, material shortages, breakdowns, quality and tooling problems assist in lowering workforce morale. If the effect of these factors is to cause in excess of 20 per cent loss, then this is certainly the scenario to be investigated. The information target is to determine which of the factors have become repetitive events.

Note: Susan Luong, the management consultant of this case study, had many standard consultancy investigation procedures and checklists that she could use. In this instance, the *five steps to managing change* were considered to be appropriate. An extract for information is given in Table 9.1. Each of the steps can be incorporated into proposals to improve the case situation.

Susan will use her skills in the outlined problem area for Wards Ltd in the next section.

Managing Change

Table 9.1 Outline of investigators' data requirements.

The five simple steps in tabular form
1. Implement controls and measures to give
(a) Labour variance
(b) Non-productive time
(c) Labour employed – overtime
(d) Output efficiency to schedule
2. Develop continuous improvement teams by
(a) Using teams of five people
(b) Focusing on specific areas
(c) Brainstorming ideas developed
(d) Actioning points for following meetings
3. Identify improvements by recognising
(a) data/information sources
(b) difficulties and 'knock-on' effects
(c) solutions
(d) costs
(e) savings

(*continued on p. 158*)

Table 9.1 (*continued*)

4. Action low-cost improvements by
   (a) Changing work practices
   (b) Simplifying material handling
   (c) Standardising skills and tooling
   (d) Introducing workforce empowerment ownership
5. Action improvements requiring investment by
   (a) Improving shop layout
   (b) Introducing new process equipment
   (c) Introducing new material-handling equipment
   (d) Improving the environment

D. Introduction to the Problems of Wards Radiator Refurbishers Ltd

Wards is a small company which provides a fast radiator repair service for vehicles ranging through large vans, lorries and articulated 44 tonnes. The company policy is to give a same-day service for radiators received before 10 a.m. and a 24-hour service for those received after 10 a.m. This is needed to suit customer requirements for maintenance and vehicle turnaround. The demand has increased over the last few years, making it necessary for more radiator repairers to be employed and trained. A father and son constitute the upper management. The father runs the commercial side of the business while the son is factory manager over the five repairers and three ancillary staff. Wards is the smallest firm in the sub-group and was purchased indirectly by taking over a group of companies when a Saltwards' competitor went bankrupt. There has always been a question mark about the company ever since, because of its size, pressure to sell it off or close it down exists. It has been profitable but cash flow is now a problem.

The layout of the factory is shown in Figure 9.4 which was not completely planned but has developed stage by stage from the original two-man operation over the last 25 years.

Output is essentially restricted by the methods of the working practice used. Decisions based on extra staff, weekend working or overtime to meet demand have not proved effective over the last year. Failure to meet delivery promises is so serious that the management accepts it has lost some potential orders. At present the capacity is achieved by five operatives working to a set target of four complete radiator repairs each per day. Hence, the best achievement expected is 20 radiator repairs per day.

Demand is firm and consistent at around 150 radiators per week (30 per day). Trying to fill the gap between supply and demand through extra hours is now self-defeating. Absenteeism is rising and some overtime refused. Some dramatic changes to the system used have to be made to resolve these problems, and time is short. In the recent past, some attempt at modifying the methods used had been made by the father. Only cosmetic changes were made, such as different kinds of incentive schemes, changes in sandblast and paint spraying equipment, and also some changes in shift patterns. All had eventually failed to make a difference after an initial favourable response when management took close interest and control of the altered situation. At this point the management decided to contact the sub-group headquarters and request assistance.

Wards were considered to be too small to have a development budget of its own, so it was included in a larger budget at sub-group Headquarters.

A positive response was given to its request because the firm was seen to be non-viable in the long term. Profit margins had diminished steadily on a monthly basis. Consequently, the Engineering Services Department were asked by Headquarters to send a consultant for three days' work, and then to report on all possible alternatives considered available, but to recommend only one.

Susan Luong was the consultant given the task of reporting to Wards and Headquarters management in about two weeks with her proposals. As the investigative area was small and involved few people, she decided not to form the usual cross-interest team but to make it a one-woman investigation using her experience.

The Consultant's Method

Basic data collection was number one priority after a 'getting to know you' period. Susan's past experience told her that company records would be unreliable or non-existent. However, output statistics were generally acceptable, also income and expenditure cash flow, but departmental and work performance figures needed to be verified. She decided to work from 'bottom up', where the delays were visible, and to establish shopfloor data on activities on an hour-by-hour basis.

An up-to-date plan of the work area was drawn (Figure 9.3) to enable her to record the radiator movements, delays and cycle times.

At the same time a Flow Process Chart was produced that generally reflected the progress of the radiators through each of the five activities (Figure 9.4), shown as follows:

D delay
⇒ transport
O operation
□ inspection
▽ storage

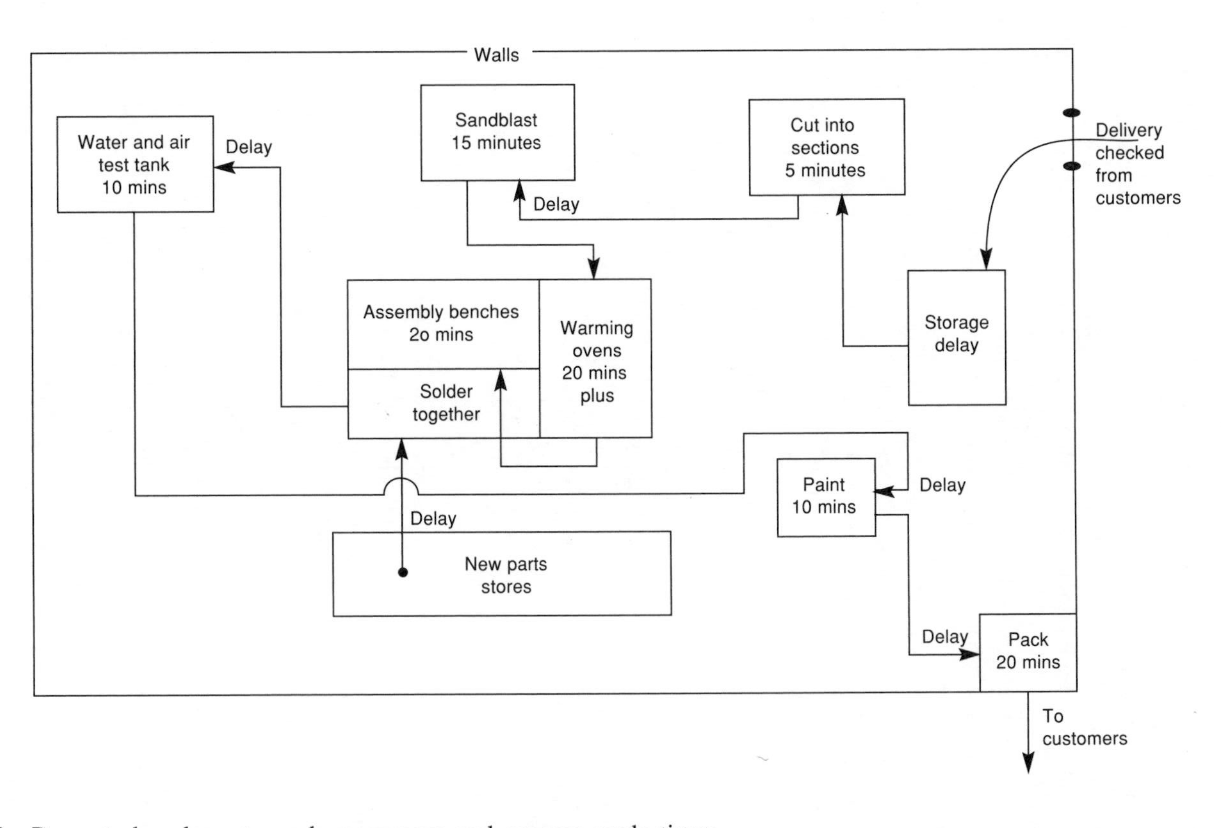

Figure 9.3 Present shop layout. work movement and process cycle times

PRESENT ✓
PROPOSED

MATERIAL MOVEMENT CHART

	SUMMARY								
		Number Pres.	Number Prop.	Man hrs. mins. Pres.	Man hrs. mins. Prop.	Move Distance in Feet	Pres.	Prop.	
Part RADIATOR									Plant.
Part No.	○	7		72		Man Power	115m		Dept. REPAIR
wt. 3 kg. Vol.	⇨	7				Crane & Hoist			Chart No. 4763
P. E. Designation Line Production	□	2		15		Truck			Sheet 1 Of 2
Process ✓	□	/				Conveyor			Chart By D COWARD
Batch Short Order	D	6		140		Total Distance Moved (m)	115		Chart Begins 1
Transport	▽	2		Unproductive Area Present	Unproductive Area Proposed	Cubic Equivalent Present	Cubic Equivalent Proposed		Chart Ends 24
Quantity/W Quantity/B ONE									Date Feb 97

	DESCRIPTION OF EVENT	Operation	Transport	Quality Check	Quantity Check	Delay	Storage	Quantity Per Load	Distance Moved (m)	Freq. Week Day Hour	Labour Force	Actual Time	Man. Hours Minutes	LOAD CHARACTERISTICS METHOD OR EQUIPMENT
1	Goods inward	○	⇨	□	□	D	▽	1	10	Hr				Hand.
2	Store / delay	○	⇨	□	□	D	▽					60m		"
3	Cut into sections	○	⇨	□	□	D	▽		10			10m		"
4	Inspect for quality	○	⇨	□	□	D	▽					5m		
5	If scrap go to stores	○	⇨	□	□	D	▽		20			15m		
6	Move to sandblast	○	⇨	□	□	D	▽		15					
7	Delay	○	⇨	□	□	D	▽					15m		
8	Sandblast top and bottom	○	⇨	□	□	D	▽					15m		
9	Move to preheat	○	⇨	□	□	D	▽		10					
10	In oven	○	⇨	□	□	D	▽					10m		
11	If reqd new core waits	○	⇨	□	□	D	▽					30m		
12	Assemble & solder	○	⇨	□	□	D	▽					15m		
13	Move to test	○	⇨	□	□	D	▽		15					
14	Wait for test	○	⇨	□	□	D	▽					10m		
15	Test radiator	○	⇨	□	□	D	▽					12m		
16	Move to paint	○	⇨	□	□	D	▽		25					
17	Wait for paint spray	○	⇨	□	□	D	▽					20m		
18	Spray with paint	○	⇨	□	□	D	▽					5m		
19	Dry paint	○	⇨	□	□	D	▽					10m		
	TOTALS													

PRESENT ✓
PROPOSED

MATERIAL MOVEMENT CHART

		SUMMARY						
Part RADIATOR		Number Pres.	Number Prop.	Man hrs. mins. Pres.	Man hrs. mins. Prop.	Move Distance in Feet	Pres.	Prop.	Plant.
Part No.	○					Man Power	115m		Dept. REPAIR
wt. 3 kg. Vol.	⇨					Crane & Hoist			Chart No. 4763
P. E. Designation	□					Truck			Sheet 2 Of 2
Line Production Process ✓	□					Conveyor			Chart By D COWARD
Batch Short Order	D					Total Distance Moved (m)	115		Chart Begins 1
Transport	▽			Unproductive Area Present	Unproductive Area Proposed	Cubic Equivalent Present	Cubic Equivalent Proposed		Chart Ends 24
Quantity/W Quantity/B ONE									Date Feb 97

	DESCRIPTION OF EVENT	Operation	Transport	Quality Check	Quantity Check	Delay	Storage	Quantity Per Load	Distance Moved (m)	Freq. WEEK DAY HOUR ✓	Labour Force	Actual Time	Man. Hours Minutes	LOAD CHARACTERISTICS METHOD OR EQUIPMENT
20	Move	○	⇨ ●	□	□	D	▽		10					
21	Delay	○	⇨	□	□	D ●	▽					5m		
22	Pack for customer	○ ●	⇨	□	□	D	▽					5m		
23	Final insp / docum.	○	⇨	□ ●	□	D	▽					10m		
24	Despatch to customer	○	⇨ ●	□	□	D	▽					30m		
25		○	⇨	□	□	D	▽							
26		○	⇨	□	□	D	▽							
27		○	⇨	□	□	D	▽					Over 4 hours PER RADIATOR		
28		○	⇨	□	□	D	▽					WORK CONTENT PER OPERATIVE 1 hr 30 m.		
29		○	⇨	□	□	D	▽							
30		○	⇨	□	□	D	▽							

Figure 9.4 Flow process chart of best performances achieved during investigation. Best performance is 5 radiators per day; 'normal' performance is 4 on average.

This gives a picture of the interruptions to flow and the cycle time of throughput plus the distance travelled.

The zig-zag pattern of movement is at its best when it only moves between operation and transport.

Throughput time was recorded for each radiator by noting the time of receipt and time of delivery. Through answering the basic questions outlined in the matrix (Figure 9.1), Susan had enough information for analysis and a means of communicating with Wards management at an early stage.

Finally it was revealed from the company records that, although the potential capacity with five operators was 20 radiators per day, the actual achievement was well below this figure. A good estimate of general performance was about 16 per day in Susan's opinion for the normal 8-hour day. Overtime at night and weekends could not close all of the gap on the present demand of 30 radiators per day. The delivery promises were being regularly broken and customer complaints were occurring.

The 24-hour service, well-advertised, was being made impossible because of the delays and the load put on the system. Management had only been partially aware of the problems, but with the new information Susan could supply, it came face to face with the detailed size of capacity shortfall. She had established the size of the problem and, hence, how much change and effort the company needed to make it viable.

The alternatives considered by Susan

A main aim of Wards' strategy was to prevent further competitors' gains, since at present it was the market leader in the region. However, to ignore the disappointed customers' needs would allow its competitors to grow and become a serious threat, so speed of change was of the essence for Wards.

Susan had to consider alternatives to increase capacity in excess of 30 radiators per day to ensure success. The strategies of 'do nothing' or 'downsize' the company to just a few regular customers were not to be evaluated at this time, Headquarters had decided.

Growth strategies considered by Susan included:

1. Operate two shifts of five operators, giving $16 \times 2 = 32$ radiators per day.
2. Employ another two or three workers and use overtime to close the output gap.
3. Subcontract the low-paying radiators and complete the high-paying radiators in-house to maximise profits.
4. Duplicate the processes that are 'bottlenecks' and employ two more repairers.
5. Move to another site and replan the radiators' work flow to reduce the present 80 m work flow and provide space for up to 10 repairers.
6. Cut down the distance each radiator moves from the present 80 m to about 25 m on the present site to allow a closer, more visual control situation.
7. Use the 'division of labour' philosophy to change work practices in a dramatic way. Make specialist operators instead of using one worker to repair the whole radiator.
8. Use more aggressive inspection of top and bottom boxes to eliminate difficult repairs. Then supply new component radiator boxes rather than a refurbished radiator. Cost to customer on these occasions could be increased but service time reduced.

Each strategy was considered and quickly evaluated. Some, necessarily, had more time and effort spent on them than others. The above ideas were not the only ideas that came under discussion because, as usual, Susan had consulted all the people involved.

The Consultant's Findings

Clearly the shopfloor managers had little reliable information concerning cycle time, delays, bottlenecks, absenteeism or individual output figures. Past performance was not important to them, only the specific problems of the day and the week ahead. The father was a 'firefighting' manager who used little planning and co-ordination skills. Often he was over-optimistic eventually: somehow everything would fall into place and he would be proven right.

Essentially, the main facts obtained by Susan Luong were based on the flow process chart (Figure 9.3) which revealed delays. This chart, which highlighted the problems of a typical radiator repair, helped Susan's discussions with management to be factual and progressive. However, this chart cannot show manpower movements, it would take a Multiple Activity Chart to reveal these activities. Susan decided not to spend time completing one. Originally the repairer and radiator moved together around the facilities, making the system have the least division of labour. Now this method had to be modified to try to counter the queues that always occurred since extra repairers had been taken on.

A quick sampling exercise also showed that each operator took different times at each operation, because some were more skilled at certain operations. Only one repairer could use each facility at a time because there was no duplication of equipment. The new method involved, as before, the radiator and repairer moving together around the workshop. However, if a queue was encountered the radiator was left at the process to be delayed while the repairer made progress with another radiator somewhere else in the workshop. This duplicity of workflow was restricted to three radiators, more often two, for each repairer. Work in progress was therefore encouraged. It resulted in repairers moving around the operations looking for available equipment. The time for queue delays is given elsewhere but it must be remembered that it is harder to save time in these queues than to exceed the normal operation times, which generally could be reduced a little but exceeded by a considerable degree.

Problems arose when queues appeared to consist of one or two repairers in length but at the side of the process where two or three extra radiators. If anyone jumped any of these informal queues, arguments followed, causing more delay and lost output. It was activities like this that restricted output to 16 radiators per day, resulting in profit loss from the missing output. There was no doubt in Susan's mind that the reactive management style and the failed delivery promises were the two main issues to be addressed.

Susan had her own specification agenda for this situation which was based on speed of output and the cost of any changes recommended. The result she wanted was a more satisfactory output meeting the schedule target and establishing adherence to a 24-hour maximum delivery performance.

E. Questions for Reader and References

Try to answer as many of these issues as possible.

1. Think about and analyse each of the eight alternatives that Susan listed, and note points for and against.
2. Can a quick method change be effected at minimum cost from any of the alternatives to meet Susan's specification agenda?
3. State the reasons for any choices made and give the expected outcomes.
4. Outline a plan to make the changes required happen, with an estimated timescale included.

Do not forget; commit ideas to paper so that they can be clearly analysed, by anyone, in a way that intangible random thoughts alone can never be.

Suggested references for further reading

Production and Operations Management, D.M. Fogarty, T.R. Hoffman and P.W. Stonebraker, South Western, USA (1989).

Production and Operations Management, A.P. Muhleman, J.S. Oakland and K. Lockyer, 6th edn, Pitman Publishing (1992).

Production Operations Management, Text and Cases, T. Hill, 2nd edn, Prentice-Hall (1990).

Work Study, R.M. Currie, edited by J.E. Faraday, 4th edn, Pitman Publishing (1977).

F. Proposals made by the Consultant

The essential points of Susan's report highlighted how to obtain an immediate improvement in output coupled with a minimum of investment. The changes should produce an improvement in cash flow so the surplus generated could be saved and used in a scientifically measured way on further improvements.

The fact that scientific data concerning time cycles and delays for all processes was not available had to be initially put aside. Investing money and time to find out the full detailed size of an already large problem would not be useful at present. A quick reasonably accurate solution dealing with the major problem would suffice. Each of the eight potential strategies was judged against the specification set by the consultant and resulted in the following tactical decisions:

Strategies
1. Duplication of effort would not work.
2. Shifts would be more expensive and a training delay is unavoidable.
3. Without detailed data, it was difficult to know the individual profit margin of each radiator repaired.

Some of the strategies were initially too expensive. Susan recommended the one strategy that met the specification in full.

This was strategy 7, using the division of labour, would cost very little and had the potential of a dramatic improvement in output.

The Strategic Proposal Outlined

The joint movement of men and radiators should be discontinued with a change in the division of labour used. Each man would be given a specific part of the repair process, not the whole. Every process would be allocated to a responsible person who would verify and monitor throughput. This is to encourage ownership and identification with any success achieved. A bonus would be that the system would become more visual in its control with the new method

Suggested specific allocations for each man:

Process 1 Collect, cut and inspect top and bottom boxes, take them to sandblast or get replacements from the stores.
Process 2 Sandblast top and bottom boxes, inspect and move to preheat furnace.
Process 3 Remove boxes from preheat, assemble and solder the three sections together; leave at water test.
Process 4 Test radiator for leaks, take to paint spray if OK or return to assembly.
Process 5 Spray with paint and leave to dry. Take dry radiator to packing station, pack and complete documentation.

From Susan's preliminary cycle time data collection she thought that a 15-minute time cycle was achievable at each process. The result would be an output of 4 radiators per hour or 32 per day. This would meet the present demand, almost overnight, and allow delivery promises to succeed. Any output problems would be readily identifiable because of the target set, 4 per hour, and ownership responsibility.

Another factor that would aid cycle time achievement was the 'learning curve' philosophy. This states that rapid improvement in performance is normal, because of repeatability, so strengthening the 32 plus radiators per day output.

Specialist operators should not themselves be delayed, only the radiators would wait at any stage. An analysis of the Flow Process Chart (Figure 9.4)

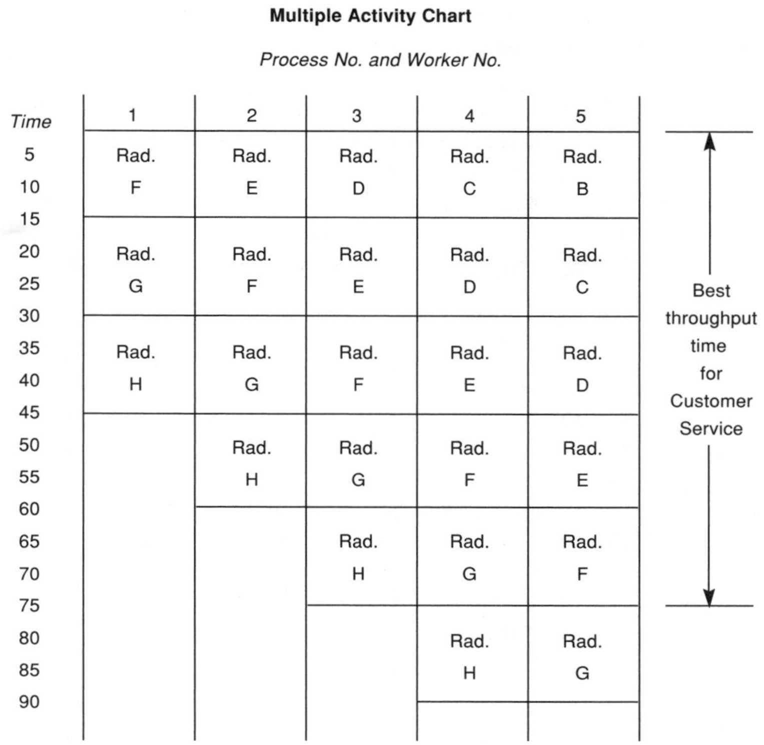

Multiple Activity Chart

Process No. and Worker No.

Time	1	2	3	4	5
5	Rad.	Rad.	Rad.	Rad.	Rad.
10	F	E	D	C	B
15					
20	Rad.	Rad.	Rad.	Rad.	Rad.
25	G	F	E	D	C
30					
35	Rad.	Rad.	Rad.	Rad.	Rad.
40	H	G	F	E	D
45					
50		Rad.	Rad.	Rad.	Rad.
55		H	G	F	E
60					
65			Rad.	Rad.	Rad.
70			H	G	F
75					
80				Rad.	Rad.
85				H	G
90					

Figure 9.5 Theoretical flow of radiators through the system to achieve target process time

showed that overall lead time would be reduced from 4 hours to under 2 hours using division of labour (see Figure 9.5 for confirmation). The planned result involves less work in progress with all workers knowing what is expected of them on an hourly basis. Difficulties are isolated to a particular process and can be reported to management for action. From the figures obtained, work balance for each worker would not be equal. However, localised process interaction decisions concerning extra help or overtime could be made effectively.

Normally it is the responsibility of the management how it decides to organise and introduce any alterations, but Susan decided to recommend the following plan.

Installation Plan

1. An initial meeting of all staff should take place to outline the need for the changes.
2. Genuine consultation concerning specific job alterations is necessary.

3. In the event of disagreement, efficiency tests of operators on equipment involved will be carried out. The better operator at each process would normally be chosen.
4. A trial period of two months, to begin one week after consultation is complete and widely agreed, was needed.
5. Regular meetings are needed to decide areas for discussion and action between men and management, firstly on a daily basis then eventually weekly, when appropriate. Management interest and commitment will be demonstrated by their close involvement with each feedback change.

Management Reaction

The father was in favour of adopting all the recommendations and starting straightaway, because he saw the need as before. However, in spite of this he was influenced against the plan by his son's reluctant attitude, who produced an opposition report which included a list of 'cannot do because' statements based on his past experiences and prejudices. Only when pressure from Headquarters was applied to the father was the plan put into action. The son was given an ultimatum which clearly said, make these changes a success or face the consequences with Headquarters. To some extent the inertia that helped produce the problems still existed during the change phase of the plan.

Outcomes

Susan was prompted by Headquarters to return and oversee the introduction of her plan. The workers eventually accepted the new proposals after one week of talks. This was mainly because they were fed up with the present system. They saw that their situation often consisted of being regularly shouted at in a climate where the job sequence had constantly to change to meet queue conditions. To be complimented on a job well done was unknown to anyone.

It was realised by the workers that they would know they were successful when they achieved the obtainable target of 4 radiators per hour on their process. The only major change that occurred came after the two-month trial. Workers wanted job rotation to be introduced on a weekly basis. Susan was happy to oversee this change to boost the morale of the workers even more.

Within two weeks of the new system being put into action, 30 radiators a day were being processed. Certain anomalies did surface and were dealt with through the meetings of workers and management. The daily meetings proved to be very useful; workers stating how they now felt part of the company instead of a pair of hands.

After the two-month job rotation trial, the system was modified to something closer to 'job swapping' between selected pairs of workers.

The sandblast process was found to be the 'bottleneck' activity and so regulated the output at a pacemaking 32 radiators per 8 hour day. An investigation to alleviate this constraint was carried out. As the equipment was old and sandblast technology had moved on, the purchase of new equipment was easily justified. Finance came from the increased profits and capacity was potentially increased to 36 radiators per day. A number of measures of performance were introduced based on each of the five processes, and targets were set for such parameters as maximum delay, process performance and the effect on costs of non-standard, non-regular radiators to exercise better control on prices.

Summary

The process of change has really just started; the new management must guard against maintaining the *status quo*, which will lead to trouble. Gradual self-financing of change started with a step difference in output brought about by the advantage of the division of labour. Theory has its place in altering methods and attitudes but each problem has to be faced in the uniqueness of different prevailing conditions.

The shopfloor manager eventually was sent on a management training course and then offered a position elsewhere in the group. This position involved a more predictable and controllable outcome.

Footnote

As a result of the obtained throughput time (Figure 9.5), the management has started talking of offering a 'while you wait' service at extra cost.

10 Capacity Shortfall? – Use Problem-Solving Techniques

Section

A. Learning Outcomes

When a problem exists it is essential to investigate with an open mind; too often people decide the solution to be adopted without enough insight into the problem.

This case study will demonstrate that:

1. Using similar methods to those that helped produce the problem is not the best way to solve it.
2. When more output capacity is required, duplication of equipment is not always the answer.
3. The fundamentals of heat transfer must not be put second to economy of scale.

4. Lack of knowledge concerning activities on the shopfloor is a clear pointer to production not being under management control.
5. Being reasonably successful should not prevent fresh approaches to increase performance.

B. Learning Objectives

After studying and analysing this case study the reader should be able to:

1. Explain the importance of correct shopfloor data collection.
2. Identify key features of problem solving.
3. Define capacity needs to allow for full customer satisfaction.
4. Understand the problems of working in an environment in which the feedback of performance is poor to both management and workers.
5. Discuss the need to build up a solution from the bottom activity and not accept the perceived wisdom of experience until it has been verified.
6. People who helped create problems are often unsuitable to be prime movers in correction activity.

C. The Basics of Problem Solving

It is easier to say 'There is not a problem or it will clear up soon', than to accept the full reality of the situation. Otherwise it is necessary to embark on a series of activities that can alleviate the trouble. This reluctance to face up to the problem is because people consider they already have enough difficulties; why look for more?

The component parts of coming to terms with the above problem identification are:

(a) it is recognised that there is a problem;
(b) the solution is not readily available;
(c) the need to solve it is clear;
(d) there is enough will to implement a solution when it is found.

What is a Problem?

'A problem can be defined as any situation in which a gap is perceived to exist between what is and what should be' (Van Gundy[2]).

As the problem situation is perceived and defined, a course of action to narrow this gap will result. Take a situation that is complex and a multifaceted problem. Analyse the following situation and decide what should be done to resolve the situation quoted.

You are travelling towards an important morning meeting with a colleague in a car. The car starts to fire erratically and finally coughs to a halt just 20 minutes away from the destination. Unfortunately it is snowing and the emergency services are likely to be fully engaged; there are 40 minutes before the start of the meeting. The car has stopped on a clearway and there are some buildings in sight. What is the problem and what should be done?

Depending on what is considered to be the problem results in that situation being analysed and solved.

What is the problem identified? Do you think it is one of the following? Does your companion agree with you? If she does not, how does this affect your actions?

1. How to get the car started?
2. How to get to the meeting on time?
3. How to find alternative transport?
4. How to get the car repaired?
5. How to get the car into a safe position?
6. How to contact the AA/RAC?
7. I wish I had a mobile phone.
8. I wish I could think up a good excuse.
9. I wish I had a more reliable car.
10. I wish it would stop snowing.
11. I wish I had been on a car maintenance course.
12. How will I get back home?
13. Have I enough money to afford alternative transport?
14. I must walk to the buildings and use a telephone.
15. I wish I had brought my wellingtons and heavy coat.

Some people are merely wishful for a solution; some take action. Many possible satisfactory courses of action are available but the one chosen will depend on how the situation is perceived. This is why sometimes the wrong problem is tackled and has little affect on the final outcome.

The problem above is difficult to resolve because the two travellers do not have enough knowledge or information to base their actions on. Data collection and analysis are a key step to obtaining a useful outcome. If the travellers had known an AA vehicle was only two miles away and would soon be passing them, or the nearest building was occupied and had a telephone etc., then their actions would then become clearer and more certain.

In most situations the key to good problem solution is data collection, followed by analysis and insight. Data collection needs careful consideration and it is often dangerous to rely on data that have already been collected. These records need challenging to check if they give the necessary indicators about the source of the problem. Investigators who collect information to their own design, such as computer programs, record sheets or video/photo-

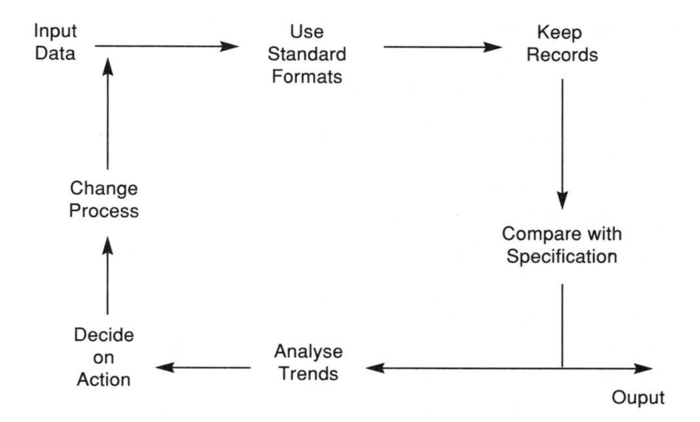

Figure 10.1 Outline of a design investigation loop

graphic evidence, will feel in control of the project. An outline of a designed investigation loop is shown in Figure 10.1. The use of a closed loop of feedback data and actioned changes is the framework on which success is based.

Reliable data collection comes from a design system that quickly and simply collects information as it becomes available. On the other hand, an incorrect instance of data collection is when a company records total percentages of scrap and rework without recording the reasons as well. A list of reasons quoted later, from experience, does not point at the 'important few' which constitute the majority of problems. Some poor managers, given a list of ten reasons for reject work, decide they can correct four of them, not realising they are only occasional faults and eliminating them has little effect on the final outcomes. The whole point of data collection is to find the size of the one or two reject categories that dominate the situation so that effort will be applied to these. Data collection needs to be by the most appropriate means. The methods used include:

1. Responsibility is given to the workers or inspectors to fill in/mark ready-prepared forms so that all questions concerning the situation are collected then.
2. Automatic data collection is used, which has the advantage of bypassing the effects of worker involvement.

Both mechanical and electronic devices can be used.

Analysis

After successful data collection the next step is analyse the results. This naturally presupposes that the analyser is fully trained and experienced.

From the analysis an expected outcome includes the following:

1. The problem is fully defined.
2. An explanation of the problem is given, built around
   (a) identification of the root causes of the problem,
   (b) place where the problem indicates itself,
   (c) when the problem happens,
   (d) how serious the problem is, and what it will cost to rectify.

After this testing and verification activity, successful alternative action can be determined.

A systematic plan must be known to all involved before the problem-solving starts. The alternative is chaotic chopping and changing, chasing the latest recorded information, resulting in an investigation that lacks form and structure.

One of the reasons complications can readily occur is because there is rarely a purely technical problem. Each problem and final solution have the extra dimension of human involvement in some way. Hence, the need to involve people, train them and win them over are very important features. Other factors include motivation, absenteeism and initiative, confirming that the recommended solution should address these issues.

Some Techniques of Problem Solving

The skills of changing problems into success have many confusing aspects; that is why they are valued by employers. Some basic techniques help investigators to solve problems. Investigation skills are directed towards funding, choice of techniques, their form and assessment of the degree of uniqueness in each situation.

The following are a few of the tools used, when required.

1. Brainstorming

A full discourse on this technique is not possible here but essentially it consists of a team of various skilled, interested people who, when confronted by a question asking for alternatives actions, allow their imaginations to 'free wheel' and respond in any way they wish.

Rules of brainstorming

1. No criticism is allowed, evaluation will come later.
2. 'Free-wheeling' is encouraged, wild ideas are welcome.
3. Quantity is wanted – the longer the list, the more chance of a radical format change.

4. Combination and improvement are encouraged, try to build on others' ideas.

Example
How many ways are there to let someone in a house with the TV on know that you are at the door? The responses will be basic, novel and sometimes ridiculous: for example, 1. Ring the door bell, 2. Use a fog horn, . . . 43. Put a brick through the window.

In exercises like this one, more than a hundred responses can be collected. As problem solving is about choosing the best alternative, brainstorming provides a suitable list. The skill is in rejecting some and developing others to achieve a new impetus to performance. This kind of technique is used to break down 'roadblocks' in people's minds, since these can stop the process of change being different and positive. It does not solve the problem but creates conditions where success is more certain.

2. Pareto curve or 80/20 rule

This is based on the fact that bias towards certain activities exists in most, if not all, situations and problems. It presupposes that some factors are more important than others. The idea of the 80/20 rule is to indicate the degree of bias that often exists. Because bias exists, the problem solver should look for this effect to gain insight into a problem.

Examples of 80/20 bias discovered

(a) 20 per cent of a warehouse's stock will equal 80 per cent of the value there.
(b) 80 per cent of maintenance problems in a vehicle fleet will come from 20 per cent of the fleet.
(c) 80 per cent of absenteeism at a company will arise from 20 per cent of the workforce.
(d) 80 per cent of rejects of a process will result from 20 per cent of the faults that occur.
(e) 80 per cent of a country's wealth is owned by 20 per cent of the population.
(f) 80 per cent of late deliveries will arise from 20 per cent of the products produced.
(g) 80 per cent of a company's sales value will be obtained from 20 per cent of its product range.

The 80/20 figures may not be exact in each situation but something close to these will be observed, for example 75/25, 85/15 or 70/30. These examples of bias are the starting point of any investigation. Which factors are dominant so they can be tackled first is the primary question?

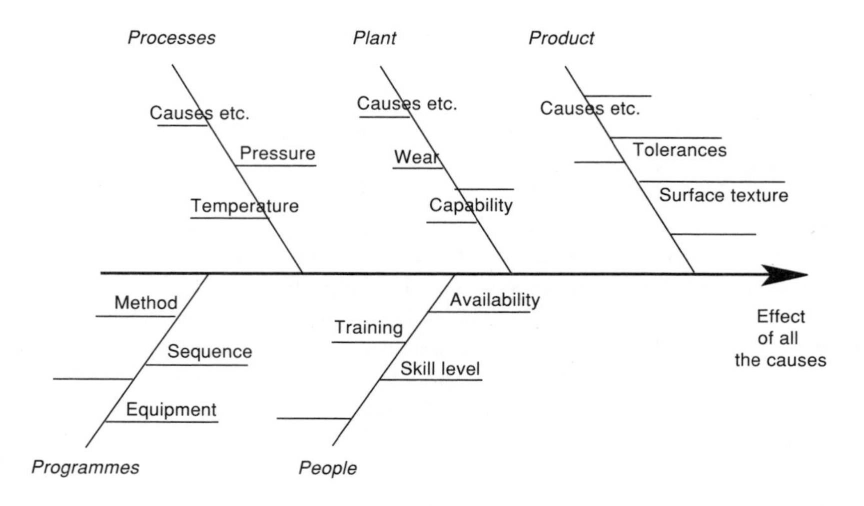

Figure 10.2 The cause and effect – Ishikawa diagram

3. Ishikawa/Herringbone diagram

Another investigation tool which links brainstorming and Pareto ideas together is the Ishikawa diagram (named after the Japanese originator), see Figure 10.2. Essentially it is a cause-and-effect diagram which has groups of potential problems listed on five main arms. The effect is shown as a horizontal arrow while the potential causes are shown under the headings of Process, Plant, Product, Programmes and People.

The 'Process' arrow would include methods of working and the environment. The 'Plant' arrow would include the equipment variables identified. A 'Product' arrow would have the design and functional needs of the process output. The 'Programmes' arrow would include control and quality tests to maintain output. Finally but not least, the 'People' arm would carry the suggestions of the potential problems from people involved in a particular case; these would include training, motivation and experience. The diagram gives hard information about the situation and its complexity, to remind all involved not to have a closed mind about possible corrective actions. Data collections and investigation of all variables should be taken.

4. Method study

Chapters 6 and 9 have demonstrated the usefulness of a scientific approach to control investigations through six major steps. One of the reasons for the repetition of this technique comes from its fundamental strengths of data collection, critical questioning and the development of alternatives that can be universally applied.

This technique can be used as the overall plan of any investigation.

Professionals of all types should make themselves familiar with method study advantages. Questioning most situations is important to gain insight.

Questioning demonstrated on a quality problem (powder coating surface defects)

1. Which product is not achieving specification in surface appearance?
2. What are the reasons for failures in surface finish? For example, runs, chips, hairs?
3. What are the problems observed?
4. Where on the product are the surface finish defects?
5. When was the problem first noted?
6. How often have the defects been detected?
7. What is the effect of the problem in £s?
8. How many products are affected?
9. How much of any one product's surface is affected.

This similar to one of the steps in method study, 'Examine', but shows a different direction of attack.

Other useful techniques

If more information about investigation techniques is required the following list could be helpful:

(a) *Quality circles* Here the direct workers take the lead after management provides preparation and training. Few companies have made a success of this system because the industrial relations it is built on are not as good as they need to be. The Japanese and some European companies have reported successes but often all the company's departments are not included in this success. It is local and patchy and comes at the end of many improvement programmes.
(b) *Operational research* This is the scientific mathematical analysis of problems, hopefully resulting in optimum answers being identified. Techniques using formulas, search and heuristic methods are used to solve problems less complex than much of real life. However, the transportation method has been successfully transferred to computer analysis to give effective answers for distribution of goods. Other methods include value analysis, network analysis, time management, quality assurance and total quality management. Consult the references listed if more information concerning these last few methods is considered useful in any future investigation.

Problem Examples Highlighted: Conflicts and Attitudes

Here an example of a poor situational circumstance is given, which you should try to resolve.

Problems in the wiring department at Spinners Ltd

Managing Director's assessment (provisional):

1. High labour turnover in the department.
2. Difficulties of recruitment.
3. Local unemployment below national average.
4. Married women operators are only supplementing family incomes for holidays, furniture etc.
5. Absenteeism is rife and such absentees are not motivated to do a good job when they do attend.

Result: Output capacity is dropping.

Investigator's assessment (initial):

1. Turnover rates are twice the area average.
2. Pay rates are compatible with the rest of the area.
3. Recruitment procedures are not particularly lax.
4. The wirer's job is unskilled and repetitive.
5. Women do not complain about the job even though it is undemanding.
6. Job satisfaction comes from the social contact between workers.
7. No negative statements about the work have been given.

Conclusion: Need to know more about the likes and dislikes of the workers.

Answers to questions set from a data collection exercise
Initial summary: The women liked the work and their fellow workers but would prefer not to work for Spinners Ltd. Everyone complained about the management but refused to be too specific.
Action: Data collection necessary. Take a week observing the activities of the wiring department.
Result obtained: This revealed supervision problems concerning insensitive treatment: shouting at workers without first obtaining the facts of any situation.

No attempts are made to correct obviously wrong decisions taken by the management on the basis that it needs to be tough and uncompromising. Often management action consisted of using 'a sledgehammer to crack a small nut'. Over-reaction was rife.

Workers' performance was measured by number of hours per week (no lateness tolerated) rather than output and quality achieved. Extravagant promises were made if the department could achieve certain levels of output, and then extra overtime was asked for to ensure success. When successful, poor rewards were given when it was obvious that the company had gained financially.

The investigator now knew what was causing people to leave or use absenteeism as an indicator of frustration. Management-induced problems like this are the hardest to resolve. Success can often only be achieved through factual documentation of many incidents to produce a pile of evidence that cannot be pushed aside. However, in the end, the investigator is 'biting the hand that feeds him' by criticising the management, including the role of the Managing Director. This is a scenario for the 'Wisdom of Solomon' to be applied.

D. Introduction to the Product Concerned

This case concerns the manufacture of *gudgeon pins*, a critical component in automotive engines. The gudgeon pin is used to attach the piston to the connecting rod. If the gudgeon pin wore excessively, a knocking noise would be the result but the engine would still function. However, if the strains on the engine pistons caused a defective gudgeon pin to fracture, the results would be catastrophic. An unattached piston/connecting rod combination would essentially 'blow up' and break the engine completely. Only the degree of damage would be unknown because damage will depend on the speed of rotation of the engine at that time.

Essentially the component outline looks like a tube about 10 mm diameter and 70 mm long. However, its manufacturing specification is narrow and demanding for this obviously critical component's reliability and safety.

Problems in production had occurred from time to time in the recent past. This product has seven different processes to complete it from the raw material. An early critical process is heat treatment. This process is called *annealing* and is used to soften the steel alloy to make it suitable to withstand the stresses and strains induced by subsequent processes.

Annealing consists of heating the gudgeon pins to a high temperature, defined easily by some as 'red hot'. After a short soak at this temperature, the component must be cooled down slowly without causing further hardness which can be induced by cooling too quickly. Verification of the annealing process is determined using a hardness testing machine, which attempts to mark the surface of a test component with a hard sharp probe. The amount of indentation created is a measure of the component hardness. A small indentation gives a higher degree of hardness.

One of the problems of heating many parts in a furnace is that each one can have a different experience of the heating/cooling cycle. This variation must be at a minimum to achieve consistent results. A further complication is because the hardness verification is made through sampling. Only 5 per cent of each furnace load is tested. When an out-of-specification gudgeon pin is found, the rest of the load is then 100 per cent hardness tested. Protection comes from this activity; poor components are eliminated from the production cycle.

Over the past 3 years, quality records have been kept but were incomplete in as much as only the 100 per cent inspection activity was noted in the style of how many had failed to reach specification. Results of samples were not recorded and also hardness values of failed components were ignored. These omissions later caused the company embarrassment when investigations were requested by their customers' quality assurance departments. This is a classic example of inspectors finding out useful information but then not saving it in full, but just the outline figures.

The Specific Problems at Pin Production Ltd

All companies that survive in effective business for over a year must have certain attributes for success and something special to offer their customers. These attributes include quality, delivery, price and/or technological developments.

Nevertheless, it also follows that a company can also have weak areas. The consequence is that weaknesses/problems need to be identified, analysed and corrected. Often the problem is only known in general outline, with specific details missing. This is the position that the top management wants tackling concerning capacity problems. Capacity needs to be increased but the present target is not often reached for gudgeon pins. Gudgeon pin output must increase from the present maximum of 35 000 per day to at least 40 000 per day inside the next two months, and this is not the full potential customer demand looking a year ahead. Of the seven processes involved in the production of gudgeon pins, all but one are capable of meeting the 40 000 per day standard. The exception is the Heat Treatment process of annealing. This process is the 'bottleneck' and sets the pace of daily output. Only one furnace is used, which is 15 years old and appears to be limited by size and the process cycle time. Annealing the gudgeon pins is a process at the forefront of quality performance.

If any gudgeon pin is overheated or soaked too long, it becomes 'soft' and cannot be easily shaped in the dies of the next operation. The required specifications cannot then be achieved. Conversely, if the gudgeon pins are not subjected to enough heat, they remain too hard and can damage the tools of the next operation. Often it is the middle of the load located in the furnace that can be too hard.

In the past, 100 per cent inspection had been used to sort the good from the bad product, but this had been discontinued after the cycle time for heat-soaking the load had been increased. This was in the hope that consistency would come from using a slower process but it still had its quality problems. However, 100 per cent hardness testing of the output was considered too expensive and slow to continue using. This inspection action was cut to sampling techniques six months ago. Middle management at the company considers it is obvious that either a new higher technological furnace is required as a straight replacement, or a further smaller furnace must be purchased to increase capacity. This smaller furnace could be new or refurbished. These furnaces could be expensive solutions as initial enquiries on cost range from £50,000 to £130,000. Customer quality assurance demands that this important component is close to the specification laid down. It is desirable that sampling inspection of hardness achieved is reliable, because of the expense of 100 per cent hardness testing.

Failures in the past had not been reported so the process appears under control. Faced by customer demands that reflect the quality image required by them, now means that quality must be organised and guaranteed. The cost of loss of confidence in the present automotive business could result in recall and replacement of thousands of engines. This means a risk of bankruptcy that the gudgeon pin company is not prepared to face. This risk is built into the latest contacts to supply gudgeon pins. Obviously, greater output puts strains on any company and puts quality assurance procedures to the test. A cocktail of difficulties can be seen to exist.

Initial Investigations by the Quality Manager

George Fairfax, the newly appointed Quality Manager, had taken over from a split management responsibility for quality. He had produced a Quality Manual and Quality Procedures to cater for the gudgeon pins production specification. He had been informed about the overall capacity demanded and understood the keenness that existed in middle management to purchase increased furnace capacity. Meetings with management resulted in requests to bring in the Engineering Services Consultancy team to clarify the furnace situation, and had resulted in a decision that George Fairfax should initially investigate and then give an interim report in two weeks.

George naturally knew about the heat treatment process, but only superficially. He now needed to obtain detailed information to gain the expertise to draw conclusions. The techniques of method study mentioned in previous chapters were once again the guiding light. As the first step of 'Select the process to be studied' had been taken, however, the best way to consider this problem is to state the reason for the investigation; namely higher output at no extra cost. This is not justification for new equipment. The second step available, 'Record the present system', was the one George concentrated on. The initial base line statistics he wanted were obtained from the gudgeon pins' steel suppliers. He asked them what the heating cycle for one

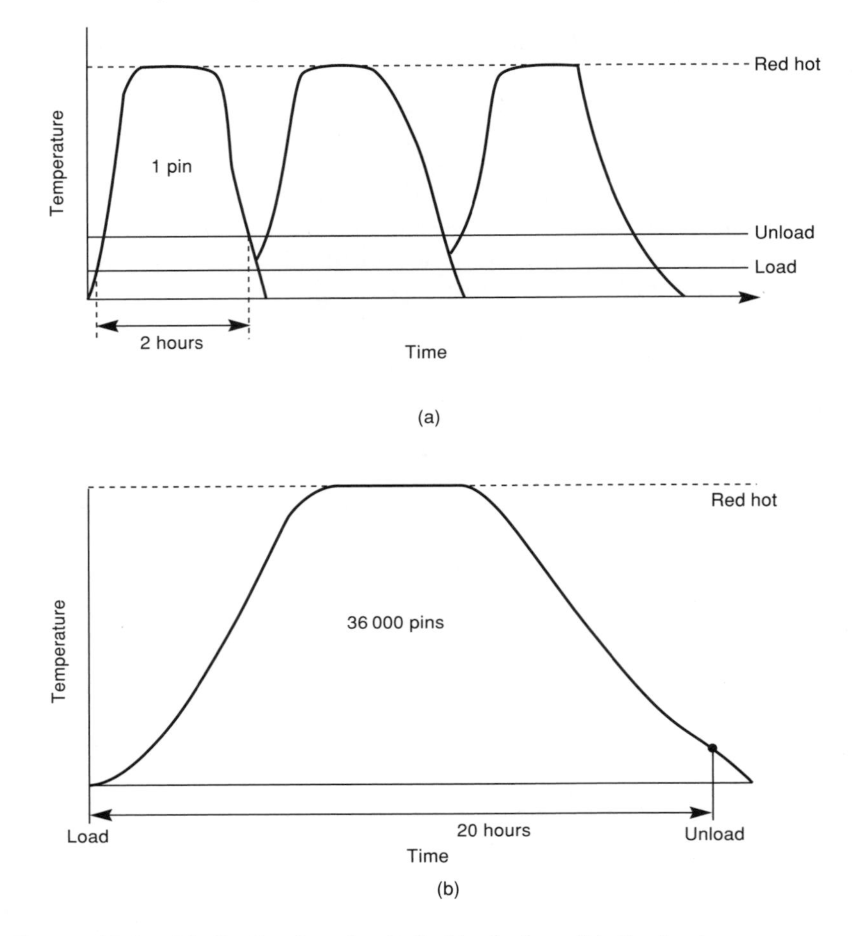

Figure 10.3 (a) Cycle time for individual pins. (b) Cycle time for full load to ensure each pin is correctly treated

gudgeon pin would be to achieve the annealing specification. Tests had shown that with a fast heat input to give the right temperature and cooling, control was easy, because for each of the individual gudgeon pins tested it was possible to have a successful 2-hour cycle (see Figure 10.3(a)). Such data were in stark contrast to the normal operational cycle time of 20 hours for thousands of gudgeon pins.

The next activity was for George to time the loading of the gudgeon pins, then the process and finally removal from the wire racks which supported the components and held them apart to allow heat penetration (Figure 10.3(b)). Specification patterns of racked loads, quantity loaded and hardness results of specific nominated sample gudgeon pins were tested at different sections of the load, that is for a range of positions – outside, inside, top, bottom and middle were nominated. George supervised the activity closely to ensure that accuracy and completeness were recorded.

Results of Interim Investigation

The specially constructed heat treatment pallets formed a 750 mm sided cube. Each layer had 200 gudgeon pins loaded on to it and there were 10 layers for each pallet. Hence, 2000 gudgeon pins were loaded in the furnace with each pallet. The furnace space was optimised by 18 pallets. Hence the full capacity was 36,000 gudgeon pins. Cycle time was set at 20 hours, so only one load was processed each working day.

Sampling tests revealed approximately 1000 were outside hardness specification, often spread in equal numbers of either too hard or too soft. Not surprisingly, the outside gudgeon pin samples were sometimes too soft and the inside gudgeon pin samples were often too hard.

An initial conclusion of George Fairfax was that the furnace could be operating overloaded and that if two furnaces were used, the second one must be able to take a larger load to counteract the proposed shortfall in load size.

Through George's questioning of long-service staff and the production control department, it was discovered that as the demand on capacity had increased over the years, so had the load and the cycle time of the annealing. The interim report gave all the above details along with a request from the investigator that the outside consultants should not be sent for at present. George was confident that, given the chance, he could resolve the problem in three weeks using the rest of the method study steps. Management acceded to his request.

E. Questions for Reader and References

1. Outline alternatives facing George Fairfax now he has the support of the management concerned.
2. Explain what course of action seems appropriate to follow the events so far carried out.
3. How should the method study steps of 'Examine the process' and 'Develop a better method' be tackled?
4. Outline the problems and constraints that are involved when adjusting a production process by running tests on actual output capacity and quality.
5. Use the graph of Figure 10.4 to give possible insights into the process that would lead to future action.

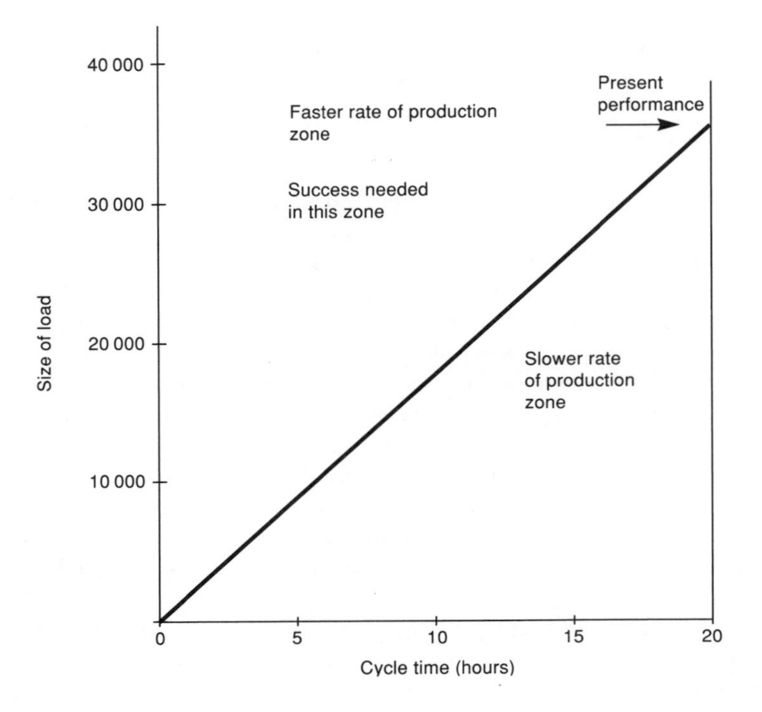

Figure 10.4 Quality Manager's idea to assist test pattern formulation

Suggested references for further reading

Justifying Investment in AMT, IEE/CIMA, Kogan Page (1992).

Production and Operations Management, D.M. Fogarty, T.R. Hoffman and P.W. Stonebraker, South Western, USA (1989).

Production and Operations Management, R. Wild, Cassell (1988).

Work Study, R.M. Currie, edited by J.E. Faraday, 4th edn, Pitman Publishing (1977).

F. Action Taken by Company

After receiving permission to complete the investigation into the heat treatment process, George set to work with urgency. He had sketched the basic formula of Figure 10.4 and expected improvements above the critical line shown. This expectation was based on the gudgeon pin test of a 2-hour cycle for the one component. If something close to 'ideal' conditions were created inside the present furnace then effective cycle time could drop.

With these plans in mind he wanted to set up a series of tests that successively lowered the load size of the gudgeon pins while the cycle time was lowered sufficiently to gain productivity. Production output demands meant he could not use the normal 5-day working week because the disruption would affect customers' orders. George submitted an oral request backed up by a memo. This asked for weekend working with two members of staff. He had already sounded out two suitable workers and received positive replies. It took a few meetings and a full presentation of his test plans finally to convince the Managing Director that the £2500 (cost fuel/wages/expenses etc.) involved was worth it. Some of this cost would be offset by output but only after extensive quality testing.

The Quality Manager's Assessment of Results

At present, each loading pallet had 2000 gudgeon pins and 18 pallets constituted a load. This situation had occurred over 4 years, ever since the demand started to grow. Successively more gudgeon pins had been loaded and the cycle time extended a little, however quality problems caused the cycle time to be extended further. The present process procedure is the end result of this activity. This *ad hoc* non-scientific procedure driven by output per day had produced these unsatisfactory outcomes. The test plan was to reverse this kind of thinking and to allow space between loaded gudgeon pins to encourage heat input and correct cooling to reduce cycle time per pallet loaded.

Plans outlined by George called for tests on the following loads, each one providing successively better rates of output per week based on increased efficiency of heat transfer:

Start position: 18 pallets of 2000 in 20 hours

Test 1	14 pallets of 2000 in	14 hours
2	10 pallets of 2000 in	9 hours
3	5 pallets of 2000 in	4 hours
4	18 pallets of 1000 in	7 hours
5	12 pallets of 1000 in	4 hours

Note: Further tests based on results obtained may be used.

The tests confirmed some of George Fairfax's ideas. Each test produced work to specification. The time available did not allow further reductions in cycle time for each load, so the optimum was not obtained. Output was systematically tested by hardness testing, taking samples from known problem areas. The most consistent results were obtained from Test 5 using the 4-hour cycle. Test 4 with the 7-hour cycle was not far behind in repeatability of hardness. When operating conditions were considered, George realised that Test 5 would require overtime or extra shifts because of the repetitive activity needed using a 4-hour cycle. It followed that Test 4, also acceptable,

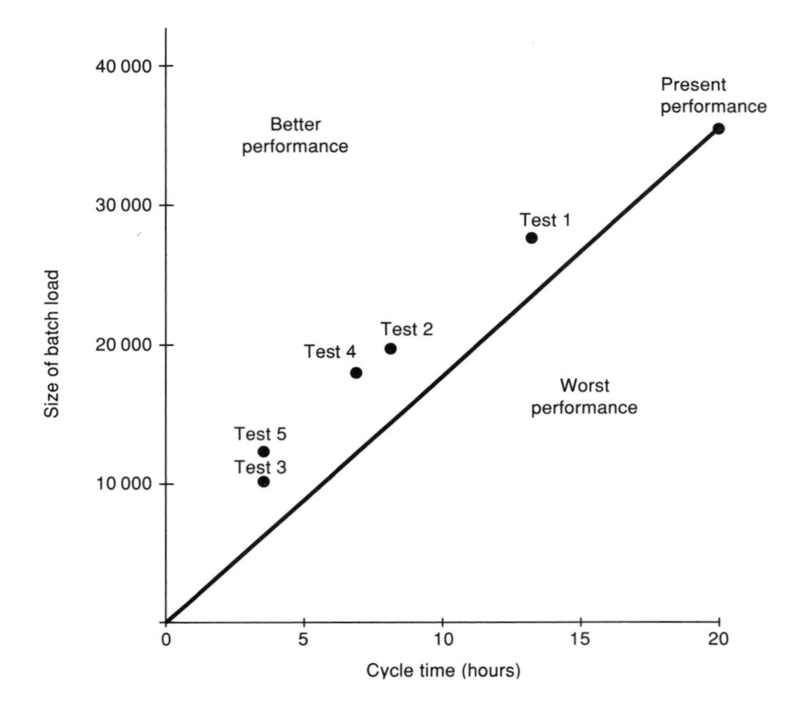

Figure 10.5 Results of output per batch time using a series of tests

had the advantage of providing the necessary two loads per day in an 8-hour manned shift because the furnace had an automatic control cycle (see Figure 10.5). The plan was to unload and load at the beginning of each shift, heat-treat the load over 7 hours and then inspect previous loads for hardness. With one hour of the manned shift available for the unload and load activity, then the process continued after the shift finished. The next day the unload/loading of pallets was done at the commencement of the shift.

All the results and findings were presented at a meeting of all management concerned and George's recommendation of Test 4 was offered. This method would allow the 36 000 gudgeon pins to be achieved as present but with the added advantages of:

1. Time was available for extra loads.
2. Only 14 hours of furnace time was used, allowing a potential further load and increasing output by 50 per cent.
3. Output quality would be improved, which is designed to please customers greatly.
4. Smaller batch sizes were able to be manufactured through the rest of the processes, reducing work in progress and also queuing.

5. No extra equipment is needed, saving at least £50,000 or more.
6. The potential exists for further savings; Test 4 results could be further adjusted by local small operational changes of the operating conditions not tested so far.

Management was pleased to agree with the recommendations and the new working pattern was instigated the following week. In this case, little preparation cost other than procedures' manual changes were required. When the extra demand for gudgeon pins came from customers, the company had a number of options available, including mixing Tests 4 and 5, to achieve the required output.

Summary of Attributes

Through identifying the real problem involved in the output capacity restriction, that is the number of components treated per hour, it was furnace over-loading that gave the key to success. Time was not wasted investigating a range of furnaces with their different options although the capacity specification of the present plant was still in doubt. Definition of the problem started from considering process efficiency and not closing down the thought processes by assuming others must have done the study correctly in the past. Establish the extent of a problem and do not consider the answer until all the facts are known.

Text Reference

2. Van Grundy, *Techniques of Structured Problem Solving*, 2nd edn, Van Nostrand Reinhold, New York, 1988.

11 The Organisation of a Factory Relocation

Section

A. Learning Outcomes

One of the difficult tasks of project management is to achieve the laid-down requirements inside the set budget and the set time allowed. To be successful requires planning and control experience, using theory but strongly balanced in favour of using simplicity. Communication difficulties are best tackled using a simple message.

This case study will demonstrate that;

1. Having a plan and people to control it cannot guarantee success; it is necessary to have 'hands on' responsibility from an identified person.
2. Budgets given are not always generous, so producing effective strategies to match the resources is essential.
3. Effective plans are made at each stage of the planning cycle. Preparation for data input to a computer and control of the computer output are all links in the same chain.

4. Processing on a computer is often necessary and the outputs of clearly printed plans are important for communication to those involved.

B. Learning Objectives

After studying and analysing this case study the reader should be able to:

1. Explain why preparation and control of plans lead to success.
2. Identify key features of effective planning and control.
3. Define the needs of project planning for a resource move.
4. Understand the problems of making a plan work and the ingredients of 'on time' and 'on cost'.
5. Differentiate between the cause and effect involved in a large range of complex activities.

C. The Basics of Organising a Project

The constraints of time and money are ever present in working life. A study of various projects over the last few years gives a list of overcost and overtime. Hence, the success of being inside both parameters is a cause of justified pride. Some of the 'failures' are familiar: Channel Tunnel, Concord, various bridges, Hospital Computer Systems, Olympic Stadiums etc. However, as they range from first-time-ever projects to repeats of other projects, excuses can always be found. Once a project is overcost and overtime it cannot be reversed, the only sensible outcome is to learn from the experience. One of the main problems faced is 'optimism', with ignorance; plans do not allow for 'contingencies' or potential problems, and budgets and timescales are not always feasible. The degree of error achieved on some of these activities can seem incredible to the general public.

A number of parts of a project have to be successful to obtain the overall benefits. The tools that are used involve *planning* and *control*. Planning for success is often difficult and is given a bad name by many. The alternative of no planning is not worth discussing. Effective action on both planning and control is required before the chance of success is established.

The Use of Critical Path Analysis

There are many good books on this subject and there is no point in duplicating their contribution. However, the thrust of the technique is often given to the middle section of the planning and control cycle. No doubt this is because of the appeal of a systematic method that gives correct solutions and is ideal for computer analysis. This thrust is fully acceptable and necessary

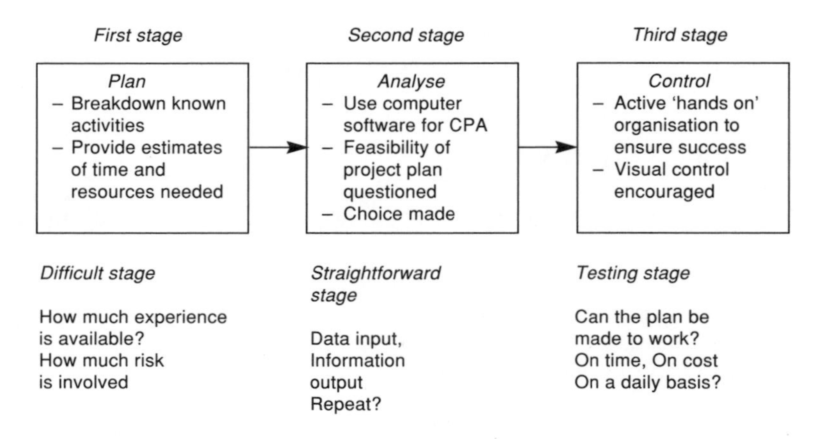

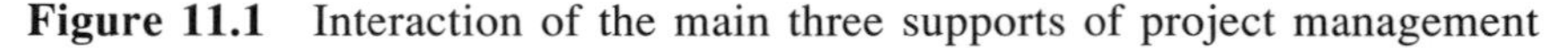

Figure 11.1 Interaction of the main three supports of project management

but only covers a part of what is required. The notes covered here will concentrate on the needs of the first section and the final third section. The three parts are shown diagrammatically in Figure 11.1 and demonstrate how the computer analysis, using CPA, is supported at the beginning and the end; these are too often ignored in text books.

The first section is potentially the most difficult. It is here that activities are identified and split into a number of sub-activities. The degree of activity breakdown is one of the keys to success. Time periods can help in these decisions, involving the use of the simple estimation rule; namely, estimate in days and error is in days, estimate in weeks and error is in weeks, estimate in hours then error is in hours etc. Depending on the overall time cycle, the size of each activity can be chosen.

The range of choice available in the breakdown of activities can be shown by a simple example.

Example activity breakdown: planting a rose bush

This could be the only activity, but it could be broken down into different levels of activity such as: obtain spade, obtain rose bush, obtain water and fertiliser. Establish the position for the hole, dig hole, fill in hole, water the plant. Tend to roses' needs, obtain stake, tie bush to stake, or even travel between shops and garden. If hundreds of rose bushes needed planting then this may justify the full breakdown of activities. If the activity is for one rose then one activity may be enough. However, considering all the component parts of planting will help identify the danger areas that could make the plan fail to meet the targets.

Critical Path Activity Components

First stage

Identify key activities noting those that can be done concurrently and those that must be done in a specific sequence.

Time estimates for each activity must then be estimated using scientific methods to gain accuracy. Normally limits inside ±10 per cent are expected. Contingency allowance should be added, depending on the level of risk of experience of the estimator. This stage is really important as it is critical in setting the foundations of the plan to be originated.

Second stage

The data established are entered into a computer program, and the network analysis programme gives an optimum solution for the data. If the use of resources and overall time quoted are satisfactory, then nothing extra is needed at this stage for the plan. However, if the answer is unsuitable in any way, then more resources and activity breakdown with more computer runs must be used until a satisfactory plan is originated. This stage is comparatively easy to use and optimum answers are readily available.

Third stage

With an acceptable plan, the degree of control necessary to provide success must be decided. Important factors include the following:

(a) A 'project champion' is needed to take over responsibility and for communication.
(b) The size of team and skills required must be specified.
(c) Timetables must be adhered to, as well as reports to watch time and cost.
(d) Willingness to change the plan in the light of experience must exist.

Thus organisational structure and control systems are the key to overall success. As long as the plan produced is feasible, then this third stage will determine the achievement of the final project outcomes concerning time and cost.

Summary of structure

The important features are, firstly, a clear activity specification giving estimates of time and resources needed. Secondly, to treat the first computer solution as just one of the alternatives possible. It can take a number of attempts to create a workable plan that gives management the choices of overall time, commitment of resources and overall cost. Projects are best conceived by a series of alternatives which allow the plan chosen to fit readily into the requirements of the situation. Alternatives come from considerations such as the following: One available worker could only carry out a sequence of activities. The rose bush is planted in a series of individual actions taking the maximum time, that is, all activity is time-summed. Two workers would allow both sequential and simultaneous actions. The rose bush can be planted with one person digging the hole, while the other obtains the rose, stake, fertiliser etc. This enables the overall cycle time to be reduced but at extra cost. Further people, if available, would cost more but cycle time is even lower.

The third and final feature is the degree of complexity involved. Risk of failure increases for a project when there is little previous knowledge in the company, or even the world, about the activities involved. North Sea oil exploration is an example where little previous experience was available. Success is more certain if the project is based on experience and some repetitive activities that create their own specialised knowledge.

Large maintenance projects often have experienced workers and repetitive activities and hence are more likely to be completed on time. The project in question should be analysed with all these considerations in mind, so that everyone concerned appreciates the problems ahead. An organisation needs to be designed that allows for the risks involved. All of the organisational aspects for projects need to be an integral part of the overall organisation; but only in an autonomous way. A leader, 'project champion', needs the authority to make decisions inside the budget constraints laid down, without the responsibility of decision making on matters such as team membership, signing orders, changing the plan or trouble shooting using company resources. Only the Managing Director, for example, should be able to ask for an explanation in the management role of accountability. No challenge to the ownership of the project should be made without hard evidence of mismanagement. The 'project champion' is the co-ordinating head of the project in which all can identify and relate to. Personalisation of project activities is essential to prevent 'buck passing' or 'postbox' management styles. Moral is higher when an interest in problems identified is guaranteed.

D. The Problems of D.G. Salt Ltd

The company was located in a multi-floor factory in need of major renovation. Everything associated with cost centres needed action: problems of component movement between processes, poor-quality equipment, low turnover of sales and output target failures, with absenteeism and labour turnover both high. Generally morale was low because of recent redundancies. It was because of this background that Group Consultants had recommended moving to a new one-level site about two miles away. A fresh start in conducive surroundings using new production methods and control systems was the only chance for survival. The reason for this opportunity being available was based on Salt's product range which had a technological base. Electronic sensors and actuators were seen as good potential money earners, provided most of the money-losing activities were eliminated.

The move needed to meet the budget and timescale, which included cash constraints and the need to limit disruption of production while the move was in progress. Other changes were necessary and these formed projects in their own right. This case study concentrates on the elements of the move and the other projects had to be integrated with it.

Specific Requirements Identified

A feasibility study was carried out before the move was agreed by the Headquarter's Management. Originally the small team set up to consider the time and resources requested £3 million to carry out all the plans outlined in its report. This estimate was considered too high and would make the move uneconomic. Keeping to this figure would only leave the one option, shut the firm down and sell off as much as possible to other firms. A second attempt involving a rethink was requested and was granted provided one of the consultants of the Engineering Services Department joined the team. This was agreed and Jim Smith was given the task of heading a new approach to the constraints involved.

These constraints included the following:

1. 300 pieces of equipment would have to be moved two miles to the new factory, which was only held on option for 6 months.
2. 100 pieces of equipment were considered the major resource, and plans of an output schedule for each would be required to maintain output.
3. The budget had now been reduced to a maximum of £1.5 million to clarify a major part of Jim's guideline constraints.
4. Use of the two-weeks' annual shutdown was a common theme mentioned by most of management as an ideal timeslot.
5. Sales were not high enough to lose any of the three main customers' business, which would signal the end of the company.
6. Some new equipment was needed to replace equipment that could not hold the quality levels necessary for success.
7. An earlier attempt to sort the 100 major machines into 'capable' and 'not capable' of meeting the quality standards set had proved a failure.
8. Refurbishing present equipment had been used as an alternative to new investment on a few specialised machines in the past. Information about local firms in the refurbishing business was well documented.
9. The move would involve seven departments and 20 different managers.
10. A move is basically a repetitive activity and hence an outline for one set of activities for a department would suit the other six.

11. Expert outside contractors would be needed, as the maintenance department at Salt's could not resource a short time project.
12. A flexible demand on manpower resources was expected and large outside removal firms could best supply the necessary resources.

It was necessary for Jim Smith and the team to address all these features for this particular move. None of the above was written in stone but the £1.5 million was to all intent fixed. One typical adjustment to the productive capacity concerned a dedicated specialised machine which was clearly in need of replacement. This machine was need for the next two years and then the product would became obsolete. It could not last this time as it was worn, outdated and unreliable. To replace it with newer, dedicated technology would cost in the order of £250 000 but the resale value after the two years would be negligible. An alternative scheme was chosen, namely six programmable flexible standard machines which could be purchased for a similar cost. In combination they were not as fast as the present dedicated machine but would be more useful to the new product range due to start in two years' time. Hence, a small cost increase was considered worth paying for the next two years so that a more appropriate output capacity would be available later. A further advantage was the resale value of flexible output machines, which would be good because of the demand for standard machines in the industry. If the product range did not require these six machines then they could be sold off either singly or in quantity. A move from dedication to flexibility in a changing market is often a good policy.

The team set to work considering the activities of moving equipment: it attempted to produce a structured plan that would form the communication link between the 20 managers and the removal subcontractors. As problems had arisen when the quality capability of the major machines was investigated, an early decision was taken to use outside expertise. A nationally respected institution was used for this and it identified two reasons for the poor capability that it discovered.

Firstly, the setting of the major machines was incorrect for over 50 per cent of the tooling. This was remedied by retraining all machine operatives away from using the bottom of the tolerance zone of the specification as the target. Instead they were taught the advantage of using the middle of the tolerance as the target.

Secondly, 15 machines were shown to be worn and incapable of achieving the necessary quality.

Quotes for both new and refurbished equipment for each of the 15 was then obtained. Along with this investment in better equipment, the quotes obtained had to include the capability indices that each machine would deliver. This information was then used to select the purchase of new or refurbishment for each of the 15 machines with the overall budget of £700,000 to guide them. Generally, expensive equipment was refurbished and the lower-priced ones purchased new.

New equipment had the advantage of overlapping with present output so customer service was guaranteed. However, the refurbishment of the present machines needed co-ordination because of their inevitable absence from the production lines. Other task forces completed their plans and the start date for the move was agreed to be Friday at the start of the Summer holidays. With a start date agreed, the team led by Jim Smith could begin considering the development of the project plan. Essential activities had to be identified and time estimates for each activity obtained. A computer package called 'PLAN' was to be used because the Engineering Services Department had prior knowledge and success with this simple planning tool. The first attempt would clarify the basic position of the company regarding cost and overall time plus the critical activities would be identified. PLAN takes activity interaction cycle times and manpower available and iterates a 'good' result. Many attempts can be made just by changing one or more pieces of data input.

E. Questions for Reader and References

The team's thoughts and actions about its task were concentrated on achieving a successful move. This was to enable all the other projects of the autonomous new company to have the chance to succeed. Moving from one place to another could not guarantee overall success but a total failure in time and cost could guarantee the end of the enterprise.

1. List the advantages and disadvantages facing the project team at the start of its efforts.
2. Outline an overall plan of identified repetitive activities to enable the management's requirements to be met.
3. What control features should be included in the project to ensure each day's work is performed correctly?

Suggested references for further reading

Critical Path Analysis and Other Project Network Techniques, K. Lockyer and J.Gordon, 5th edn, Pitman Publishing (1991).

Production and Operations Management, D.M. Fogarty, T.R. Hoffmann and P.W. Stonebraker, South Western, USA (1989).

Production and Operations Management, A.P. Muhlemann, J.S. Oakland and K. Lockyer, 6th edn, Pitman Publishing (1992).

Production and Operations Management, R. Wild, Cassell (1988).

Production/Operations Management, Text and Cases, T. Hill, 2nd edn, Prentice-Hall (1990).

F. The Proposals Accepted by the Company

The team started with a simple calculation based on a two-week move:

> 10 working days (14 possible), 100 major machines to move and 200 ancillary pieces of equipment
>
> Result: 10 major pieces per day
> 20 ancillary pieces per day

The move for each of the above would involve the following major tasks:

1. disconnection of service;
2. movement through multi-level factory;
3. load on to lorry;
4. transport;
5. unload and move to prepared position;
6. bed machine support and connect the necessary service and commission.

This movement load was considered too high and the proposal was modified to request an extension to 30 working days. A consequence of this was the need for a plan to keep output up to demand while the equipment was split between the two sites.

For scheduling purposes, the activities chosen were not the six listed above. The following three were considered more adaptable to a day's activity:

1. disconnect services and move to loading bay;
2. transport to new site, unload and position;
3. bed in and level, connect services and commission.

The reason for choosing these three was based on geographical lines. Each activity occurs in a different place. It would be easy to identify the planning success by simple checks at the two sites regarding the equipment's position. Corrective action required would be from a completely visual input. Instead of giving a time for each of the three activities, to enable the work load to be assessed, it was decided to regulate the number of pieces of equipment for each day depending on the degree of difficulty expected. This gave the advantage of a daily planning scheme based on position. A target of *X* hours per machine was therefore not quoted to prevent arguments about time for individual pieces of equipment. Each of the seven departments was given a time slot for its move and a separate plan controlling each piece of equipment to be moved on a particular day. The overall Gantt chart for the seven departments is given in Figure 11.2, which marks the days and weeks for each.

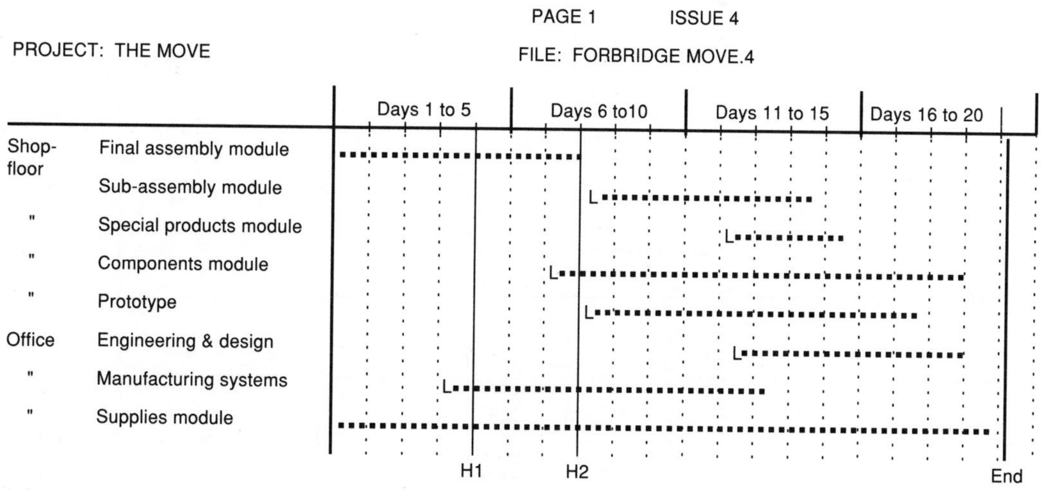

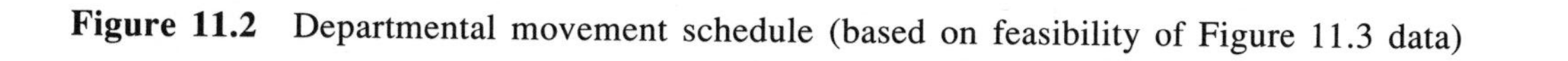

Figure 11.2 Departmental movement schedule (based on feasibility of Figure 11.3 data)

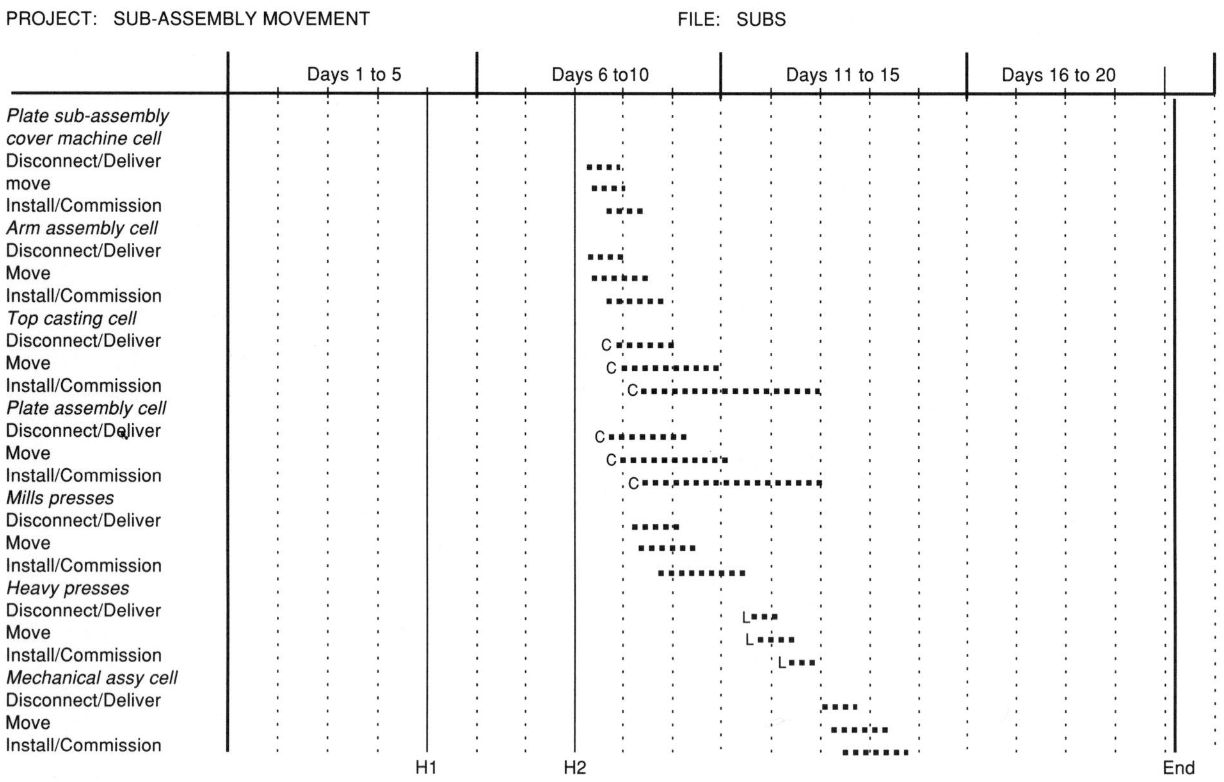

PROJECT: SUB-ASSEMBLY MOVEMENT

FILE: SUBS

Days 1 to 5

Days 6 to10

Days 11 to 15

Days 16 to 20

Plate sub-assembly
cover machine cell
Disconnect/Deliver
move
Install/Commission
Arm assembly cell
Disconnect/Deliver
Move
Install/Commission
Top casting cell
Disconnect/Deliver
Move
Install/Commission
Plate assembly cell
Disconnect/Deliver
Move
Install/Commission
Mills presses
Disconnect/Deliver
Move
Install/Commission
Heavy presses
Disconnect/Deliver
Move
Install/Commission
Mechanical assy cell
Disconnect/Deliver
Move
Install/Commission

H1

H2

End

Figure 11.3 Gantt chart showing examples of manufacturing cell scheduled movements

DISCONNECT & DELIVER TO TRANSPORT **MAIN COMPONENTS** **Issue No. 7**

MOVE TO FORDBRIDGE

Disconnect and deliver to Despatch bay the following equipment:

Monday 20 July Project day 14

Locn.	M/C No.	Description	240 V	415 V	air	water	extract
K5	DT497	Slack & Parr Quick Tap		*	*		
K5	A3808	Index auto		*	*		
K5	FH1638	Wickman Bush & Ream		*	*		
K5	A3924	Index Auto		*	*		
K5	D5060	Wickman Special Driller		*	*		

Tuesday 21 July Project day 15

Locn	M/C No.	Description	240 V	415 V	air	water	extract
G2	A3886	Wickman Auto 5/8 * 6		*	*		
G2	A3887	Wickman Auto 5/8 * 6		*	*		
G2	A3904	Wickman Auto 5/8 * 6		*	*		
G2	A3928	Wickman Auto 1 3/4 * 6		*	*		
G2	A3839	Wickman Auto 1 3/4 * 6		*	*		
G2	A3825	Wickman Auto 1 3/4 * 6		*	*		
G2	TO883	Fox Riddle		*	*		

Wednesday 22 July Project day 16

Locn	M/C No.	Description	240 V	415 V	air	water	extract
G2	A36118	CVA Single Auto		*	*		
G2	A3346	BSA 98 Single Auto		*	*		

Figure 11.4 Examples of a typical daily schedule for a piece of equipment

A more detailed Gantt chart (computer generated) for one department, shown in Figure 11.3, identifies each manufacturing cell's movement timetable. The ancillary pieces of equipment had a special team allocated to move them from set lists of equipment, scheduled for each particular week.

Figure 11.4 gives further detailed lists for the major pieces of equipment and gives specific information concerning Plant No. and Location with services required. The number of pieces of equipment to be moved in any one day was determined from estimates based on a working day from 0830 to 1600 hours. An agreement with the subcontractors allowed for working between 1600 and 1800 hours to ensure each day's target was achieved. This gave the contingency allowance necessary to cover for the unexpected, from whatever source. As an extra safeguard, weekend working would not be planned into the move but was available in the event of massive problems. Finally the plan, as outlined, needed control. The visual side has already been quoted and this was supplemented by three graduate engineers who

Index